The flowering plants of South Africa vol. 1

By

I. B. Pole Evans

PUBLISHED BY: 2024 by BTB Publishing

ISBN : 978-1-63652-368-2

THE FLOWERING PLANTS OF SOUTH AFRICA

VOL. 1

I. B. POLE EVANS

CONTENTS

Preface...1

Plate 1..5

Plate 2..8

Plate 3...11

Plate 4...15

Plate 5...18

Plate 6...21

Plates 7 And 8...25

Plate 9...29

Plate 10...32

Plate 11...35

Plate 12...39

Plate 13...41

Plate 14...44

Plate 15...47

Plate 16...50

Plate 17...53

Plate 18...56

Plate 19...59

Plate 20...62

Plate 21...65

Plate 22...68

Plate 23...71

Plate 24...74
Plate 25...77
Plate 26...80
Plate 27...83
Plate 28...86
Plate 29...90
Plate 30...93
Plate 31...96
Plate 32...99
Plate 33...102
Plate 34...105
Plate 35...108
Plate 36...111
Plate 37...114
Plate 38...117
Plate 39...120
Plate 40...123

PREFACE

THE cultivation of South African plants in Europe dates back to early times.

Indeed, it may safely be assumed that it was in vogue soon after the Dutch settlement at the Cape, for Holland during the 16th and 17th centuries held first place in European horticulture. Her cities even vied with one another in the establishment of gardens of exotic plants, many of which came from the Cape.

These treasures created such interest and attracted such attention that Cape plants soon became the fashion and object of envy throughout Europe. Collectors were specially despatched to these shores for the purpose of hunting out and securing their botanical wealth.

Evidence also is not lacking that the cultivation of indigenous plants was carried out at the Cape prior to 1700.

Be this as it may, little remains to-day in South Africa to do credit to the past preservation and cultivation of our native flora.

In recent years, however, considerable interest has again sprung up in this direction; in fact, it is rapidly becoming fashionable to have a rockery of aloes, vijgies, and other succulents as one of the chief adjuncts to the garden.

Apart from these—perhaps better-known plants—there are many beautiful flowering herbs, shrubs and trees of the veld, which might with advantage be grown in our gardens and around our homes.

It is with the object of bringing these gems of nature to the notice of the public that this publication is offered.

A work of this kind is of necessity a costly undertaking, and its future existence and ultimate success will depend largely on the support which it receives at the hands of the public.

The publication of the present volume has only been made possible through the interest and keenness of a South African lady, whose love for her country and its natural beauties has been the means of procuring the necessary funds for the initiation of the work.

It is proposed to issue this publication as an illustrated serial, much on the same lines as the well-known Curtis's *Botanical Magazine*, and for imitating which no apology need be tendered.

Should the publication be the means of stimulating further interest in the study and cultivation of our indigenous plants amongst the rising generation, the desire and object of its promoters will be achieved.

Living plants suitable for illustration, plants of economic value, or plants of general interest, will always be gladly received and welcomed by the Editor.

As regards the illustrations the Editor has been most fortunate in being able to place the work of that skilful artist, Miss K. A. Lansdell, before the public, while the descriptions have been prepared by Dr. E. Percy Phillips, Botanist in charge of the National Herbarium, to both of whom it is a pleasure to express one's special thanks for the trouble and care which they have taken.

For the information of those of our readers who have not been fortunate enough to visit our country or our inland capital, it may be added that the illustration on our cover represents a glimpse of the magnificent Union Buildings at Pretoria, under whose shadow this work is being prepared, and on whose site the plants here figured are grown.

It has been the Editor's privilege and good fortune to see a comparatively bare kopje converted in the course of a few years into the site of a grand and stately building surrounded with many of the country's most beautiful and interesting herbs and shrubs.

The illustration depicts such characteristic plants as the arborescent *Aloe Marlothii*, Berg.; the handsome *Aloe Wickensii*, Pole Evans (in the centre); *Aloe aculeata*, Pole Evans; the neat *Aloe Peglerae*, Schönland; *Cotyledon orbiculata*, Linn.; *Encephalartos Altensteinii*, Lehm; and some arborescent euphorbias.

Although most of the plants just mentioned are typical of the vegetation of the Transvaal, an endeavour will be made in our magazine to depict in each volume as far as possible an equal number of floral representatives from all the Provinces.

As the authority for colour nomenclature, *Colour Standards and Colour Nomenclature*, by R. Ridgway, Washington, 1912, has been adopted.

I. B. POLE EVANS.

Pretoria, 1920.

AGAPANTHUS UMBELLATUS, L'HERIT.

1.

K.A. Lansdell del.

AGAPANTHUS UMBELLATUS, L'HERIT.

PLATE 1.

AGAPANTHUS UMBELLATUS.
Cape Province, Natal, Orange Free State, Basutoland, and *Transvaal.*

LILIACEAE. TRIBE ALLIOIDEAE.

AGAPANTHUS, *L'Herit; Benth. et Hook. f. Gen. Plant.*
Agapanthus umbellatus, *L'Herit. Sert. Angl.* 17; *Fl.*

This well-known garden plant, commonly called the "Agapanthus" or "Blue Lily," was introduced into cultivation in England from the Cape as far back as 1692, and was figured by Commelin of Amsterdam in 1697.

In South Africa the plant is characteristic of the vegetation found on high mountain ranges. It usually occurs on well-drained slopes.

It is a herbaceous plant with a tuberous rootstock from which 6-10 broad strap-shaped leaves arise. The lower stalk is about a metre high, and bears an umbel of 20-50 handsome blue flowers.

DESCRIPTION:—*Rootstock* short, tuberous, with fleshy cylindric roots. *Leaves* dark green, 6-10, basal, 20-60 cm. long, 2-4 cm. broad, strap-shaped, obtuse, glabrous. *Peduncle* O·25-1 metre high, terete, glabrous. *Inflorescence* a many-flowered umbel. *Spathe-valves* 2·5 cm. long, 2 cm. broad at the base, ovate, acuminate, withering in the mature inflorescence. *Floral-bracts* 2·7 cm. long, linear. *Pedicels* about 6 cm. long, terete, jointed at the apex. *Flowers* blue; perianth-tube 1·7 cm. long, 6 mm. in diameter; lobes 2-5 cm. long, 9 mm. broad, oblanceolate, obtuse. *Stamens* inserted at the throat of the perianth-tube; filaments 2·5-3 cm. long, arcuate; anthers oblong. *Fruit* a triquetrous capsule.

F.P.S.A., 1920.

ALOE GLOBULIGEMMA, I B POLE EVANS.

2.

K.A. Lansdell del.

ALOE GLOBULIGEMMA, I. B. POLE EVANS.

PLATE 2.

ALOE GLOBULIGEMMA.
Transvaal.

LILIACEAE. TRIBE ALOINEAE.

ALOE, *Linn.*; *Benth. et Hook. f. Gen. Plant.*

Aloe globuligemma, *Pole Evans in Trans. Roy. Soc. S. Africa,*

This remarkable Aloe was collected by Messrs. Wickens and Pienaar in M'Phathlele's Location in the Pietersburg District during January, 1914. Specimens brought to the Botanical Laboratories, Pretoria, flowered during July and August of the same year.

In M'Phathlele's Location the plant occurs in vast numbers in a very gregarious manner on the open sandy plains. In fact, it is not uncommon to find long, continuous belts of thickly crowded plants extending for two or three hundred yards in length. The plant is typical of the Low Veld and the river valleys which run from the Zoutpansberg into the Limpopo basin.

At first the racemes are furnished with widely separated spherical to globular flower-buds which develop with considerable slowness.

The unopened flowers are rich nopal red (R.C.S.), tinged with green at the tips. When open, the flowers become a sulphur-yellow (R.C.S.).

Our illustration was made from a plant in the Aloe collection at the Botanical Laboratories, Pretoria.

DESCRIPTION:—A succulent, stemless plant. *Leaves* 16-23 in a dense rosette, glaucous, erect-spreading, 45-50 cm. long, 8-9 cm. broad at the base, lanceolate-ensiform, acuminate, acute and recurved at the apex, unspotted, somewhat flat at the base and canaliculate above, with cartilaginous wavy and toothed margins; teeth pale brown and at right angles to the margins, 1·5-2 mm. long, and about 8-9 mm. apart, deltoid, recurved. *Inflorescence* a panicle, with 5-7 spreading horizontal to oblique branches with a few small deltoid-acute empty bracts at the base. *Peduncle* 06-1 metre high, stout, glaucous, naked. *Racemes* densely flowered, 22-40 cm. long. *Flowers* secund, all pointing towards the centre of the inflorescence and at the same time slightly deflexed; young buds distinctly globular; in open flowers nopal-red (R.C.S.), green at the tips; mature flowers sulphur-yellow (R.C.S.) and tinged with red towards the base. *Floral-bracts* reflexed, 5-6 mm. long, ovate-cuspidate, scariose, pellucid. *Pedicels* recurved, 3-4 mm. long. *Perianth* 25 mm. long, cylindrical-ventricose; outer segments free for 15-17 mm., obtuse and recurved at the apex, 3-5-veined; inner segments obtuse, recurved, tipped with auburn (R.C.S.) at the apex, with 3 inner veins. *Stamens* exserted for 11 mm.; filaments slightly recurved, the exposed portion chestnut-brown (R.C.S.) to black. *Anthers* mars-orange (R.C.S.). *Style* pale sulphur-yellow (R.C.S.), stout, recurved. *Capsule* shortly stipitate, 23 mm. long, 13 mm. in diameter, oblong-ovoid.

PLATE 2.—Fig. 1, plant much reduced; Fig. 2, lower part of spike; Fig. 3, apex of spike.

F.P.S.A., 1920.

ARCTOTIS DECURRENS, JACQ.

K A Lansdell del.

3.

K. A. Lansdell del.

ARCTOTIS DECURRENS, JACQ.

PLATE 3.

ARCTOTIS FOSTERI. [A]

Clanwilliam.

Compositae. Tribe Arctotideae.

Arctotis, *Linn.*; *Benth. et Hook. f. Gen. Plant.*

Arctotis Fosteri, *N.E. Br.*

Arctotus *Fosteri*; *Herbacea*, 60-90 cm. alta. *Folia* inferiora lyrato-pinnatisecta, 30-40 cm. longa, 7-9 cm. lata, longe petiolata, ambitu obovato-oblonga, lobis lateralibus utrinque 3-4, oblongis subacutis dentatis, lobo terminali latissime ovato obtuso grandidentato; folia superiora sessilia, lanceolata, acuta, subintegra; omnia supra parce pubescentia, subtus albo-lanata. *Pedunculi* 17-19 cm. longi, striati, pubescentes. *Capitula* 7-9 cm. diametro. *Involucri bracteae* exteriores ovatae, subulato-caudatae, virides; interiores oblongo-obovatae, obtusae, apice membranaceae, rubescentes. *Radii flores* acuti, albi vel carnei, subtus purpurei, quisque basi maculo nigro et aurantiaco instructi. *Disci flores* nigro-brunnei, antheribus luteia. *Pappi squamae* lanceolatae, acutae vel subobtusae. *Ovarium* villosum.—N. E. Brown.

Cape Province: Clanwilliam Division, near Clanwilliam, *Foster.*

This handsome species of *Arctotis* was raised in the Garden of the Division of Botany, Pretoria, from seed received in 1916 from Mr. C. Foster, of Clanwilliam, after whom I have much pleasure in naming it. The large size of the flower head and the delicate colouring of the rays make it worthy of a place in all South African Gardens. At Pretoria it flowers freely and has set mature seed.

DESCRIPTION:—A herbaceous perennial 60-90 cm. high. *Leaves* many, lyrate; lower leaves 30-40 cm. long, 7-9 cm. broad, lanceolate in outline, obtuse, produced at the base into a long petiole, sparsely pubescent on the upper surface, white-woolly on the lower surface; leaf-lobes about 4 cm. long, 1-5·2 cm. broad, oblong, obtuse, with toothed margins; the terminal lobe much larger, otherwise similar; petiole up to 16 cm. long, flat on the upper surface, convex beneath, sparsely glandular-pilose; upper leaves sessile, lanceolate, acute, sparsely pilose, with more or less entire margins. *Peduncles* pale in colour at the base, gradually becoming indian purple (R.C.S.) towards the apex, 17-19 cm. long, terete, striate, pilose, the hairs becoming reddish and denser towards the apex. *Capitulum* solitary, 7-9 cm. in diameter when fully expanded. *Involucral bracts* many-seriate; the outer green, ovate, with a hairy subulate appendage; the inner reddish and membranous at the apex. *Receptacle* slightly convex. *Ray-florets* female. *Corolla* limb white or shrimp-pink (R.C.S.) above, with a golden-yellow and purple spot near the base, and eugenia red to vandyke red (R.C.S.) beneath. *Pappus* of several membranous scales as long as the corolla-tube. *Ovary* with a dense tuft of basal hairs. *Disc-florets* hermaphrodite. *Corolla-tube* 4-5 mm. long, campanulate above, cylindrical below; lobes lanceolate with black tips. *Pappus* of several membranous scales about half length of the corolla-tube. *Ovary* villous, with a dense basal tuft of hairs. *Fruit* villous, crowned with the persistent pappus scales.—E. PERCY PHILLIPS.

PLATE 3.—Fig. 1, portion of ray-floret; Fig. 2, pappus-scale of ray-floret; Fig. 3, disc-floret; Fig. 4, pappus-scale of disc-floret; Fig. 5, apex of style showing the stigmas; Fig. 6, fruit; Fig. 7, involucral-bracts (all enlarged).

F.P.S.A., 1920.

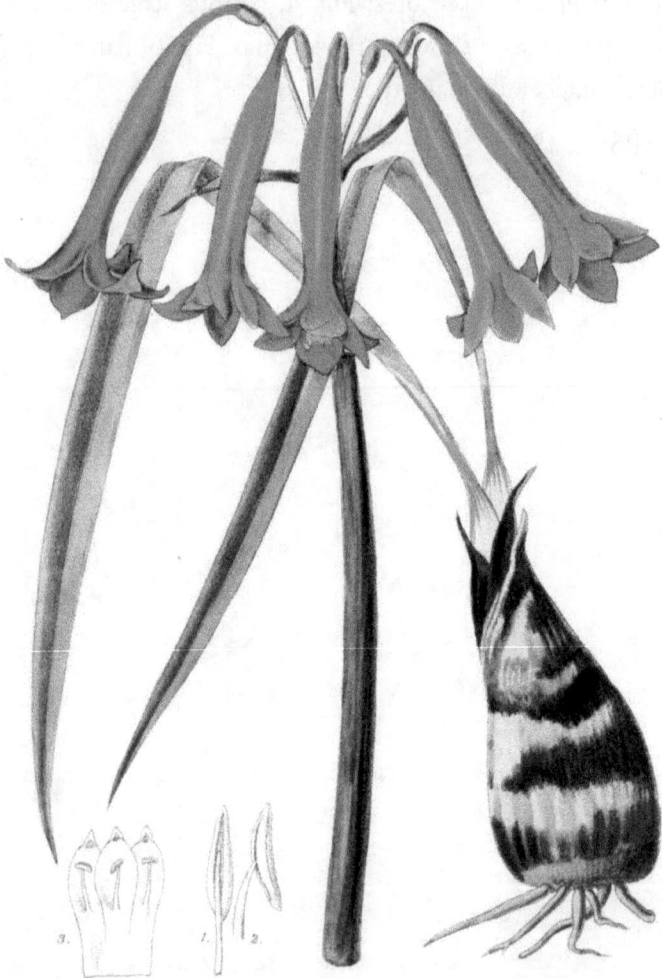

4.

K. A. Lansdell del.

CYRTANTHUS ANGUSTIFOLIUS, AIT.

PLATE 4.

AMARYLLIDACEAE. TRIBE AMARYLLEAE.

CYRTANTHUS, *Ait.*; *Benth. et Hook. f. Hook. f. Gen. Plant.*

Cyrtanthus contractus, *N.E. Br.*

CYRTANTHUS *contractus*; *Bulbus* 4-6 cm. diametro, ovoideus, brunneus. *Folia* 2-3, adscendentia, 30-50 cm. longa, 8-12 mm. lata, linearia, longe et acutissime acuminata, basi in petiolum teretem angustata, glabra. *Pedunculus* 18 cm. vel ultra longus, ad 1 cm. diametro, teres, fistularis, glaber, purpurascens, ad apicem 4-10-florus. *Bracteae* 4·5-5 cm. longae, basi 8 mm. lati, lineari-attenuatae, membranaceae. *Pedicelli* adacendentes, 2-4 cm. longi, rubri. *Perianthium* nutante, coccineum; tubus 5-6 cm. longus, ad medium leviter ventricosum et circa 8 mm. diametro, basi longe et valde contractus et circa 1·5-2 mm. diametro; lobi 1·3 to 1·5 cm. longi, 7 mm. lati, elliptico-ovati, acuti. *Stamina* perianthii lobis breviora; antherae luteae. *Ovarium* 1 cm. longum, ovoideum. *Stylus* inclusus, stigmatibus recurvis. *Semina* complanata, atrata.—N. E. BROWN.

Transvaal: on kopjes near Mooi Plaates farm, in the vicinity of Pretoria, *Miss J. Stuart.*

The specimens from which our drawing was made were collected by Miss J. Stuart of Pretoria, on the slopes of stony kopjes near the farm "Mooi Plaates," about 5 miles out of Pretoria. During the spring months the plant is very conspicuous on the kopjes when it flowers freely, usually after the grass has been burnt off, and from this it derives its common name of "Fire Lily."

DESCRIPTION:—*Bulb* 4-6 cm. in diameter, ovoid; tunics brown, membranous. *Leaves* 2-3, contemporary with or appearing after the flowers, 30-50 cm. long, 0·8-1·2 cm. broad, linear, acuminate, acute, tapering to a terete petiole above the flattened base, glabrous. *Peduncle* vandyke red (R.C.S.), 18 cm. long, lengthening in the fruit, 10 mm. in diameter, terete, hollow, glabrous. *Spathe-valves* 4·5-5 cm. long, 8 mm. broad near the base, linear, acute, membranous, glabrous. *Inflorescence*, a 4-10-flowered umbel. *Flowers* pendulous, scarlet-red to carmine (R.C.S.), faintly scented; pedicels 2-4 cm. long, strawberry-pink (R.C.S.), jointed at the apex; perianth-tube 5-6 cm. long, tubular, narrowed at the base, with an inflated portion about the middle; lobes 1·3-1·5 cm. long, 7 mm. broad, ovate-ellipsoid, acute, with a small tuft of white hairs on the inner surface just below the apex. *Stamens* inserted just below the throat, a little shorter than the perianth-lobes; anthers chrome-yellow (R.C.S.), oblong. *Ovary* 1 cm. long, 0·5 cm. in diameter, ovoid; style about the length of the perianth-tube, included; stigmas recurved. *Seeds* black, flattened.—E. PERCY PHILLIPS.

PLATE 4.—Figs. 1 and 2, anthers back and side view; Fig. 3, apex of the perianth, showing the tufts of hairs.

F.P.S.A., 1920.

GERBERA JAMESONI, BOLUS.

5.

K. A. Lansdell del.

GERBERA JAMESONI, BOLUS.

PLATE 5.

COMPOSITAE. TRIBE MUTISIACEAE.

GERBERA, *Gronov.*; *Benth. et Hook. f. Gen. Plant.*
Gerbera Jamesoni, *Bolus; Gard. Chron., 1889.*

This plant, the "Barberton Daisy," has attracted much attention in recent years as an addition to the herbaceous garden.

It was first discovered in the Transvaal by the collector Rehmann about 1878, and later by the Hon. R. Jameson on the mountains round Barberton. In 1889 it flowered at Kew Gardens and was illustrated in the *Gardener's Chronicle* for that year. Shortly afterwards a coloured plate and description appeared in the *Botanical Magazine* (t. 7087). The specific name was proposed by the late Dr. Bolus, who himself collected the species at Barberton.

A mass of these plants in full bloom is very striking, the scarlet rays standing out in strong contrast to the green leaves.

The specimen from which the present illustration was made flowered at the Botanical Laboratories, Pretoria.

DESCRIPTION:—*Rootstock* perennial. *Leaves* basal, numerous, 22-45 cm. long, 5-10 cm. broad, somewhat oblong in general outline, deeply lobed, with the terminal lobe broadly ovate, acute, pubescent on both surfaces, especially on the veins beneath; petiole 25-40 cm. long, terete, pilose, tinged with red at the base. *Peduncle* 25-40 cm. long, terete, pilose, bearing a solitary capitulum.

Capitulum 8-10 cm. in diameter when fully expanded. *Involucral-bracts* about 3-seriate, 1-1·5 cm. long, lanceolate, acuminate, acute, woolly. *Receptacle* flat, naked. *Ray-florets* female, scarlet to spectrum-red (R.C.S.); lower limb represented by 2 linear strongly reflexed lobes. *Ovary* glandular-pubescent. *Disc-florets* hermaphrodite. *Corolla* bilabiate; tube 7 mm. long, cylindrical, glabrous; anterior limb of 2 linear recurved lobes; posterior limb 3-toothed, spreading. *Ovary* 6 mm. long, cylindrical, glandular-pubescent; *stigma* bifid.

PLATE 5.—Fig. 1, ray-floret; Fig. 2, disc-floret; Fig. 3, apex of style, showing the two stigmas.

F.P.S.A., 1920.

6.

K. A. Lansdell del.

GLADIOLUS PSITACCINUS, H.K., VAR. COOPERI, B. K. R.

PLATE 6.

GLADIOLUS PSITTACINUS, var. COOPERI.
Basutoland and *Transvaal.*

IRIDACEAE. TRIBE IXIEAE.

GLADIOLUS, *Linn.*; *Benth. et Hook. f. Gen. Plant.*
Gladiolus psittacinus, *Hook f. var.* **Cooperi**, *Bkr. Bot. Mag. t. 6202; Fl.*

This striking species of *Gladiolus* belongs to the section *Dracocephali* of the genus which contains some seven species all characterised by having the upper segments hooded. It is a favourite with cultivators, and has also been taken up by nurserymen who grow blooms for sale. Like a great many of our beautiful native species, it was left to Kew Gardens to bring the species to public notice.

Corms of this *Gladiolus* were brought to England by Mr. Thomas Cooper, who collected in South Africa for Mr. Wilson Saunders. It first flowered at Kew in 1872, when it was described and figured.

The species is easy of cultivation, and flowers at Pretoria in November. Apart from the interest it is to gardeners generally, it should be of special interest to breeders, as there seems little doubt that some very fine hybrids could be raised from this species.

DESCRIPTION:—A tall plant O·9-1·2 metres high. *Corm* red, 2·5-6 cm. in diameter, globose; tunics chartaceous, broad, ultimately breaking up into fibres. *Leaves* 6-8, 60-90 cm. long, 2-3 cm. broad, ensiform, acute, slightly narrowed at the base, glabrous, rigid. *Inflorescence* a lax 5-10-flowered spike, 20 cm. or more long.

Spathe-valves 5-9 cm. long, oblong-lanceolate, sub-acute, kildare-green (R.C.S.). *Flowers* large, hooded, lemon-yellow (R.C.S.), striped with scarlet-red (R.C.S.). *Perianth-tube* curved, 5-6 cm. long, trumpet-shaped, glabrous; three upper lobes forming a hood 2-2·5 cm. broad, ovate or obovate, acute, with a distinct claw; the posterior lobe crimson without, lighter in colour within, smaller than the other two upper lobes; the three lower lobes smaller than the upper lobes and strongly reflexed; the anterior lobe 3-4 cm. long, O·5-2 cm. broad, broadly-lanceolate, acute, lemon-yellow (R.C.S.) in the lower half; the lateral lower lobes 2-2·5 cm. long, 0·4-0·5 cm. broad, lanceolate, acute. *Stamens* inserted at the base of the perianth-tube, shorter than the upper lobes, arcuate. *Ovary* ellipsoid; style slightly longer than the stamens, arcuate; stigmas terete, pubescent on the stigmatic surface. *Capsule* 2-2·5 cm. long, ellipsoid. *Seeds* discoid.

PLATE 6.—Fig. 1, plant much reduced; Fig. 2, apex of style, showing the three stigmas; Figs. 3 and 4, back and side view of stamens.

F.P.S.A., 1920.

K.A.Lansdell del

LEUCADENDRON STOKOEI PHILLIPS

7.

K. A. Lansdell del.

LEUCADENDRON STOKOEI, PHILLIPS.

LEUCADENDRON STOKOEI, PHILLIPS.

8.

K. A. Lansdell del.

LEUCADENDRON STOKOEI, PHILLIPS.

PLATES 7 AND 8.

LEUCADENDRON STOKOEI.
Cape Province.

PROTEACEAE. TRIBE PROTEEAE.

LEUCADENDRON, *R. Br.*; *Benth. et Hook. f. Gen. Plant.*
Leucadendron Stokoei, Phillips sp. nov.

LEUCADENDRON *Stokoei*; *Rami* et ramuli glabri. *Folia* 7-8·5 cm.
longa, 1-2·1 cm. lata, oblonga vel oblongo-lanceolata, apice obtu-
sa, basi paullo angustata, glabra. *Inflorescentia* ♂ 2 cm. longa, 2·5
cm. lata; bractae involucri 1·1 cm. longae, apice obtusae, glabrae
viscidaeque, eximae reflexae; bractae floris 1 mm. longae, lanceolatae, acuminatae, apice subacutae, basi villosae. *Inflorescentia* ♀ 2·5
cm. longa, 1·7 cm. lata; bractae involucri reflexae; bractae floris 6
mm. longae, 1·1 cm. latae, apice obtusae, sericeae. *Fructus* 7 mm.
longus, 6 mm. latus, obovatus, anguste alatus.

Cape Province: Caledon Division, Standford, near Caledon,
Stokoe in National Herbarium.

The remarkable group of South African plants—the *Proteaceae*—still continues to yield interesting novelties, though it was
monographed as recently as 1910. This is especially true of the
genus *Leucadendron*, on which much work remains to be done.

The specimens from which our illustration was made were
collected by Mr. T. P. Stokoe in the Klein River Mountains at
Sinkerhausgat, near Standford, in the Caledon Division. Mr.

Stokoe has made some interesting discoveries in this region, amongst which was this new *Leucadendron* which he forwarded to the Division of Botany in September, 1918. It is quite distinct from any other species of *Leucadendron*, inasmuch as the male inflorescence is surrounded by large bracts giving it the appearance of a *Protea*.

DESCRIPTION:—*Branches* and branchlets glabrous. *Leaves* 7-8·5 cm. long, 1-2·1 cm. broad, oblong or oblong-lanceolate, obtuse, bluntly mucronate, slightly narrowed at the base, glabrous, very finely glandular when seen under a lens; leaves surrounding the inflorescence longer. *Male inflorescence* hidden by the upper leaves and quite surrounded by brown bracts, shortly peduncled, 2 cm. long, 2·5 cm. in diameter; the inner bracts 1·1 cm. long, oblong, obtuse, glabrous, viscid, longer or as long as the inflorescence; the outer situated on the short peduncle, viscid and reflexed; receptacle 7 mm. long, 8 mm. broad, subglobose. *Floral bracts* 1 mm. long, lanceolate, acuminate, subacute, villous at the base. *Perianth* 2 mm. long, glabrous. *Stigma* clavate, much thicker than the style. *Female inflorescence* hidden by the upper leaves, very shortly stalked, 2·5 cm. long, 1·7 cm. broad, the short peduncle bearing brown reflexed bracts; receptacle 1·6 cm. long, 3 mm. broad, cylindrical. *Floral bracts* 6 mm. long, 1·1 cm. broad, transversely oblong, villous above. *Fruiting head* 3·5-4 cm. long, 3·5 cm. in diameter; scales 1·5 cm. long, 1·2 cm. broad, suborbicular, slightly narrowed at the base, tomentose without, except near the apex. *Fruits* 7 mm. long, 6 mm. broad, obovate in outline, flat on one side, convex on the other, honeycombed, with a narrow membranous wing.

PLATE 7.—Fig. 1, male inflorescence; Fig. 2, longitudinal section of male inflorescence.

PLATE 8.—Fig. 1, young female inflorescence; Fig. 2, longitudinal section of female inflorescence; Fig. 3, fruiting head; Fig. 4, floral bract × 4; Fig. 5, fruit × 4.

F.P.S.A., 1920.

9.

K. A. Lansdell del.
TULBAGHIA VIOLACEA PARV.

PLATE 9.

TULBAGHIA VIOLACEA.
Cape Province, Natal.

LILIACEAE. TRIBE ALLIEAE.

TULBAGHIA, *Linn.*; *Benth. et Hook. f. Gen. Plant.*
Tulbaghia violacea, *Harv. Bot. Mag. t. 3555. Fl. Cap.*

This beautiful little *Tulbaghia* has a special interest attached to it, inasmuch as it flowered in Baron von Ludwig's garden at the Cape of Good Hope, and was there illustrated and described by Harvey, who sent his description and drawing to the *Botanical Magazine* for publication. In this respect it differs from most of the other Cape plants, which were described and figured from specimens grown in Europe.

The present illustration was made from specimens cultivated at the Botanic Gardens, Durban, Natal, from tubers presented by Mrs. Todd of Pietermaritzburg. Notwithstanding the unpleasant odour of garlic, the species is worth cultivation. It is commonly known as the "Wild Garlic."

DESCRIPTION:—*Rootstock* tuberous. *Leaves* crowded; basal leaves rudimentary and membranous; upper leaves 6-10, erect, 20-30 cm. long, 5-8 mm. broad, linear, acute, glabrous, concave on the upper surface, rounded beneath. *Peduncles* erect, 30-60 cm. long, terete. *Inflorescence* a 10-20-flowered umbel. *Spathe-valves* 2-2·5 cm. long, 5 mm. broad at the base, ovate-linear, acuminate, acute, membranous. *Pedicles* 2·5-4 cm. long, terete. *Flowers* pale ampare-purple to light haryense-violet (R.C.S.). *Perianth-tube*

1·1-1·5 cm. long, 4 mm. in diameter, cylindrical, slightly inflated at the base; lobes 1-1·1 cm. long, 3-5 mm. broad, elliptic or lanceolate, acute or obtuse. *Corona-lobes* 3, 1·5 mm. long, opposite the inner segments. *Stamens* subsessile, in 2 whorls about halfway down the perianth-tube; anthers sub-globose. *Ovary* sessile, sub-globose; style 2 mm. long, thick; stigma capitate.

PLATE 9.—Fig. 1, perianth laid open to show corona lobes and stamens × 1½.

F.P.S.A., 1920.

K A Lansdell del

RICHARDIA ANGUSTILOBA, SCHOTT

10.

K. A. Lansdell del.

RICHARDIA ANGUSTILOBA, SCHOTT.

PLATE 10.

AROIDEAE. TRIBE PHILODENDREAE.

RICHARDIA, *Kunth.*; *Benth. et Hook. f. Gen. Plant.*
Richardia angustiloba, *Schott in Journ. Bot.* 1865, 35; *Fl.*

The specimen figured was first mentioned in the *Gardener's Chronicle*, 1892, as *Calla Pentlandii*, and two years later in the same publication was again brought to notice as *Richardia Pentlandii*, under which name it was described and figured in the *Botanical Magazine*. Mr. N. E. Brown, who monographed the genus for the *Flora Capensis*, considers it the same as *Richardia angustiloba* which was described as early as 1865. It was introduced into cultivation by Mr. R. Whyte, Pentland House, Lee, who raised flowers in 1892, and exhibited it at a meeting of the Royal Horticultural Society and was awarded a first-class certificate.

Specimens of the tubers were taken to Kew by Mr. E. E. Galpin in May, 1892.

Our present illustration was made from plants flowered by Mr. H. H. Wickens, Officer in Charge of the Union Buildings Garden, Pretoria.

DESCRIPTION:—*Leaves* dark green, unspotted; petiole 30-60 cm. long, terete, glabrous; lamina 22-30 cm. long, 11 cm. broad at the widest part, ovate, acute, and produced into a filiform appendage at the apex 1·2 cm. long, sagittate at the base, glabrous, with the midrib prominent beneath and channelled above. *Peduncle* over 1 metre high, longer than the leaves, terete, glabrous. *Spathe* gamboge-yellow, dark purple at the base inside, 10-14 cm. long, 4-4·5 cm. broad across the middle, 7-10 cm. across the mouth, loosely convolute for two-thirds of its length, then expanding into a broad, nearly horizontal limb produced into a subulate tip 1·5 cm. long and with recurved margins. *Spadix* yellow, 4·5 cm. long, cylindrical. *Ovaries* with subsessile stigmas, pale greenish-white. *Staminodia* none; anthers yellow. *Calla Pentlandii, Gard. Chron.* 1892, p. 124; *Richardia Pentlandii, Gard. Chron.* 1894, p. 590; *Bot. Mag.* t. 7397.

PLATE 10.—Fig. 1, plant much reduced; Fig. 2, spathe removed to show the spadix.

F.P.S.A., 1920.

11.

K. A. Lansdell del.

FREESIA REFRACTA, KLATT.

PLATE 11.

FREESIA Sparrmannii var. FLAVA.
Cape Province.

Iridaceae. Tribe Ixieae.

Freesia, *Klatt*; *Benth. et Hook. f. Gen. Plant.*
Freesia Sparrmannii, *N.E. Br.* var. **flava,** *N.E. Br.*
Gladiolus Sparrmanni, *Thunb. in Kongl. Vet. Acad.*
Handlingar, 1814, p. 189, t. 9A, and *Fl.*

According to the *Flora Capensis* the only species in the genus *Freesia* is *F. refracta*, Klatt, which is a native of the eastern districts of Albany, Bathurst, etc., and is characterised by having (among other characters) the slender lower part of the perianth-tube shorter than the upper broader part and not more than twice as long as the bracts. But there are at least three other species found in other regions that distinctly differ in habit or in the tube of the flower or in both. One of them collected by Burchell in Bechuanaland and at present undescribed, has a very long tube. Another is a plant found in the coast districts of Swellendam, Riversdale, Ladismith, etc., figured and described by Thunberg under the name of *Gladiolus Sparrmanni*, upon which I found the species *Freesia Sparrmannii*. The reference to this figure is omitted by Schultes in his edition of Thunberg's *Fl. Cap.*, and is not quoted by Baker, but it accurately agrees with the plant Zeyher collected along the Buffeljagts River in Swellendam Division and distributed under No. 4027. It conspicuously differs from *F. refracta* by the very much longer slender part to the perianth tube, and although Thunberg's plant and that collected by Zeyher have purplish-tinted flowers, I place the plant here figured as a yellow variety of it, because I find that the late P. MacOwan, in a letter preserved at Kew, gives the following particulars concerning this species, which he also considers distinct from *F. refracta*. He writes: "All along the coast from Cape Point towards Agulhas, notably near Mossel Bay, the other *Freesia* grows wild. I have never seen it in my Eastern Province peregrinations.... Its colour varies very much, from pale golden daffodil tint to pure white, and is either with or without purplish stains on the outside of the perianth-segments. Here, at the Hort. C.B.S., we paid much attention to this lovely bulb, grew it year after year, roguing out all the yellow and purple-stained individuals and sowing the whitest. This is the '*Freesia refracta alba*' of gardens."

This note gives the origin of *F. refracta* var. *alba*, Baker, *Handb. Irid.* p. 167, which should now be called *F. Sparrmannii* var. *alba*, for it certainly is not the same as the true *F. refracta*, and Thunberg's original name must be upheld.

The plant here figured is doubtless the pale golden form mentioned by MacOwan, and it differs from the yellow-flowered *F. xanthospila* by the very long slender part of its perianth-tube.—N. E. Brown.

Our illustration was made from specimens grown in the Gardens of the Division of Botany from bulbs presented by Mr. J. Shand, of Ladismith, Cape Province.

Description:—*Corm* about 4 cm. long, 3 cm. in diameter, produced into a short neck and densely covered with fibres. *Leaves* basal, 6-8 cm. long, ·5-1 cm. broad, acute, somewhat sheathing at the base, glabrous. *Peduncle* 9·5 cm. long, with the upper portion bent at a right angle. *Spathe-valves* 1 cm. long, ovate, subacuminate, acute, membranous in the upper portion. *Perianth-tube* 5·2 cm. long, 1·2 cm. in diameter above, campanulate in the upper portion and becoming slenderly tubular in the lower half, yellow; lobes 1·2 cm. long, 1 cm. broad, ovate-oblong, or subrotund, rounded above, yellow. *Style* 5-6 cm. long, filiform, 6-lobed; lobes 5 mm. long, linear, somewhat spathulate at the apex.—E. PERCY PHILLIPS.

PLATE 11.—Fig. 1, anther; Fig. 2, style arms.

F.P.S.A., 1921.

CRASSULA FALCATA, WILLD.

12.

K. A. Lansdell del.

CRASSULA FALCATA, WILLD.

PLATE 12.

CRASSULA FALCATA.
Cape Province.

CRASSULACEAE.

CRASSULA, *Linn.*; *Benth. et Hook. f. Gen. Plant.* **Crassula falcata,**
Wendland, Bot. Beobachtungen, (1798), Willd. Enum. (1809); Fl.

Among rock plants there are few which equal this fine *Crassula* for brilliant colouring. It is easy to propagate and flowers freely. The species is common in the Eastern Province, and is found in flower during the month of June. The specimen from which our illustration is made was collected by Mr. P. J. Pienaar at Grahamstown and flowered in the Gardens of the Division of Botany.

DESCRIPTION:—*Stem* succulent, 30-55 cm. high, simple. *Leaves* connate at the base, fleshy, 6-9 cm. long, 1·5-2·5 cm. broad, decreasing in size upwards, obliquely falcate, obtuse, glaucous. *Peduncle* reddish in colour. *Inflorescence* a dense trichotomous cyme. *Calyx-lobes* 3 mm. long, ovate or oblong, obtuse, canescent. *Petals* 1 cm. long, connate at the base; lobes linear-lanceolate, subobtuse. *Stamens* nearly as long as the petals. *Styles* 5, subulate. *Squamae* minute.

PLATE 12.—Fig. 1, carpels and squamae × 5.

F.P.S.A., 1921.

K.A. Lansdell del.

CLIVIA MINIATA. HEGEL.

13.

K.A. Lansdell del.

CLIVIA MINIATA. HEGEL.

PLATE 13.

AMARYLLIDACEAE. TRIBE AMARYLLEAE.

CLIVIA, *Lindl.*; *Benth. et Hook. f. Gen. Plant.*

Clivia miniata, *Regel, Gartenflora♀, 1864, t. 434;*

Fl.Imantophyllum (?) *miniatum, Hook. Bot. Mag.* t. 4783.

This species, indigenous to Natal, represents only one of many of our native plants, which have been brought to the notice of horticulturists by English Nurserymen. A living plant was exhibited at a meeting of the Horticultural Society in February 1854 by Messrs. Backhouse, who imported the plant from Natal. The specimen from which our illustration was made was collected by Miss K. A. Lansdell at Ifafa on the South Coast of Natal. The species is a shade lover, and is usually found flowering in the shelter of rocks and trees. The size and number of the flowers have been much improved by cultivation, and several hybrids have been raised from the species. The flowers may vary in colour from a red to a yellowish-red.

DESCRIPTION:—*Rootstock* a fleshy rhizome, 1·5-2 cm. in diameter, with numerous fleshy cylindrical roots. *Leaves* many, 40-50 cm. long, 5-6·5 cm. broad, strap-shaped, acute, slightly narrowed at the base, the leaf bases forming a distinct swelling just above the rhizome, glabrous, bright green. *Peduncles* shorter than the leaves, compressed, sharply 2-edged. *Inflorescence* a 12-20-flowered umbel. *Spathe-valves* 4 cm. long, 7-8 mm. broad, ovate-oblong, membranous. *Floral bracts* 2·5 cm. long, linear. *Flowers* erect. *Perianth* divided almost to the base; tube about 5 mm. long; segments 5-7 cm. long; the inner 1·1 om. and the outer 1·8-2·1 cm. broad at the widest part, oblanceolate; the inner emarginate; the outer minutely thickened at the apex; all obtuse, gradually narrowed to a claw; bright red, with white margins at the lower half. *Stamens* included; filaments compressed; anthers linear, versatile. *Ovary* 5-6 mm. long; ellipsoid, bluntly 3-angled; style slender, as long as the perianth; stigma trifid, sometimes bifid. *Fruit* a bright red berry, globose, 1·5 cm. in diameter. *Seeds* 1 or few, subglobose.

PLATE 13.—Fig. 1, section of peduncle; Fig. 2, bract; Fig. 3, transverse section of ovary; Fig. 4, style and stigmas; Fig. 5, fruit.

F.P.S.A., 1921.

14.

K. A. Lansdell del.

GARDENIA GLOBOSA, THUNB.

PLATE 14.

GARDENIA GLOBOSA.
Cape Province, Natal.

RUBIACEAE. TRIBE GARDENIEAE.

GARDENIA, *Linn.*; *Benth. et Hook. f. Gen. Plant.*
Gardenia globosa, *Hochst. in Flora, 1842,*
Bot. Mag. t. 4791; Harv. Thes. t. 5; Fl.

This handsome plant is a shrub or sometimes becomes a small tree, and is without doubt one of our finest native flowering shrubs. It is common in Natal, where it flowers in early spring and summer. The large fragrant bell-shaped flowers are produced in great profusion and give to the plant a very striking appearance. The species has been known to European cultivation for over sixty years, but is usually grown in the greenhouse. It is frequently seen in gardens in Durban, Natal, and specimens have been grown in Queens Park, East London, but the plant has not received the attention from South African horticulturists which it deserves.

Our illustration was made from specimens collected by Miss K. A. Lansdell in the Stella bush near Durban, Natal. The native name is "Isi-Qoba."

DESCRIPTION:—A *shrub* or small tree. *Branches* with dark-coloured bark, glabrous. *Leaves* opposite; petioles 3-5 mm. long; blade 5-15 cm. long, 2-3.5 cm. broad, lanceolate or sometimes oblanceolate, obtuse or acute, gradually tapering to the base, entire, with a prominent reddish mid-rib beneath, glabrous; stipules about one-third of the length of the petiole, ovate, acuminate, minutely pubescent, soon deciduous. *Flowers* terminal, axillary or clustered. *Pedicels* 1-2 mm. long, minutely pubescent. *Calyx* 3-4 mm. long, minutely pubescent and glandular without, silky within; tube campanulate; lobes acute. *Corolla* white, usually with 5 faint pink lines within, which may become darker near the base and broader on the lobes, sometimes spotted; tube 2-5 cm. long, 1-8 cm. in diameter above, campanulate, suddenly constricted and narrowed above the calyx, minutely pubescent without, densely tomentose within; lobes spreading, half as long as the tube. *Anthers* linear. *Ovary* 1-celled, with numerous ovules; stigmas white or pink. *Fruit* a brown berry, crowned with the persistent calyx-lobes, many seeded. *Seeds* minute, immersed in the fleshy parietal placentas.

PLATE 14.—Fig. 1, style arms; Fig. 2, fruit.

F.P.S.A., 1921.

RICHARDIA REHMANNI, ENGL

15.

K. A. Lansdell del.

RICHARDIA REHMANNI. ENGL.

PLATE 15.

RICHARDIA REHMANNI.
Natal, Transvaal, Swaziland.

AROIDEAE. TRIBE PHILODENDREAE.

RICHARDIA, *Kunth*; *Benth. et Hook. f. Gen. Plant.*
Richardia Rehmanni, *N.E. Br. in Gard.*
Chron. 1888, Bot. Mag. t. 7436; Fl.

This species was first collected by the traveller Rehmann and described by Engler in 1883 as *Zantedeschia Rehmanni*. Among English horticulturists the plant attracted a lot of attention, and was several times referred to in the *Gardener's Chronicle*. The chief attraction to cultivators is the deep red colour of the spathes, but when grown in English gardens and also in its native climate, the colour varies considerably. Dr. Medley Wood notes that at the Natal Herbarium, Durban, the original deep red colour returned to the plants after being cultivated fifteen years. This loss of colour, however, does not appear to be constant among plants which flowered for the first time at the Division of Botany Gardens, Pretoria, from tubers which were sent by S. G. Marwick, Esq., Assistant Commissioner, Hlatikulu, Swaziland. In these the colour ranged from pale pink to deep red. After fertilization, however, and during the formation of the fruits the colour gradually fades from the spathes and they become green. The species was introduced into England by Mr. R. W. Adlam of Natal, who sent tubers to the Cambridge Botanic Gardens. The leaves vary from a uniform green to green with white markings, or green with darker green markings.

Our illustration was made from specimens cultivated at the Natal Herbarium, Durban, Natal.

DESCRIPTION:—*Plant* about 0·5 metre high. *Leaves* 3-5, the lower reduced to mere sheaths; petiole 15-30 cm. long, deeply channelled down the face, rounded on the back, stem-clasping at the base; blade 40-60 cm. long, 6-8 cm. broad, lanceolate, acute, with a subulate point, narrowed at the base into the petiole, entire, with undulating margins, and the mid-rib prominent beneath, dark green, sometimes with white, sometimes with green markings, shining. *Peduncle* shorter than the leaves, terete, glabrous, olive-green. *Spathe* 10-15 cm. long, with a tube 4-5 cm. long and 1·8-2 cm. in diameter, with an ovate acuminate limb, varying in colour from almost white to a deep rose or aster purple (R.C.S.) in the upper portion, greenish-yellow below, without a dark blotch round the base of the spadix. *Spadix* stout, with male flowers on the upper half and female flowers on the lower half. *Ovary* glabrous; stigma sessile. *Fruit* a berry. *Seeds* subglobose.

PLATE 15.—Fig. 1, plant, reduced; Fig. 2, spadix; Fig. 3, ovary.

F.P.S.A., 1921.

16.

K. A. Lansdell del.

ADENIUM MULTIFLORUM, KLOTZ.

PLATE 16.

ADENIUM MULTIFLORUM.
Transvaal, Zululand, Portuguese East Africa.

APOCYNACEAE. TRIBE ECHITIDEAE.

ADENIUM, *R. & S.*; *Benth. et Hook. f. Gen. Plant.*
Adenium multiflorum, *Klotzch in Peters,*
Reise Mossamb. Bot. p. 279, t. xliv., Fl.

The specimen from which our illustration was made is growing in the Gardens of the Division of Botany, Pretoria, and was presented by Mr. A. E. Antrobus, Cloud's End, Louis Trichard, in the Zoutpansberg District. The flowers appeared in September before the leaves, which only made their appearance the following month. The species does very well on a rockery, and when it flowers is a very pleasing sight.

DESCRIPTION:—A plant with a very large tuber just below the ground-level and from which the branches spring. *Branches* more or less succulent, glabrous. *Leaves* appearing after the flowers, sub-sessile, 3·5-9 cm. long, 1·5-6·5 cm. broad, obovate, obtuse, narrowed to the base, dark green and very shiny above, pale green and dull beneath, with the mid-rib and lateral veins distinct above, the mid-rib alone prominent beneath. *Inflorescence* cymose, up to 5-flowered, terminal. *Sepals* lanceolate, pilose. *Corolla-tube* about 3 cm. long, 1·2 cm. broad above, tubular below, pilose without and within on the broadened portion; lobes, 1·7-2·5 cm. long, about 1 cm. broad, elliptic-oblong, or obovate, shortly acuminate, acute, with crinkled edges, usually sparsely pubescent on the upper portion, pink, with dark red margins. *Anthers* densely villous.

PLATE 16.—Fig. 1, plant, reduced; Fig. 2, leaf; Fig. 3, calyx; Fig. 4, stamen; Fig. 5, pistil.

F.P.S.A., 1921.

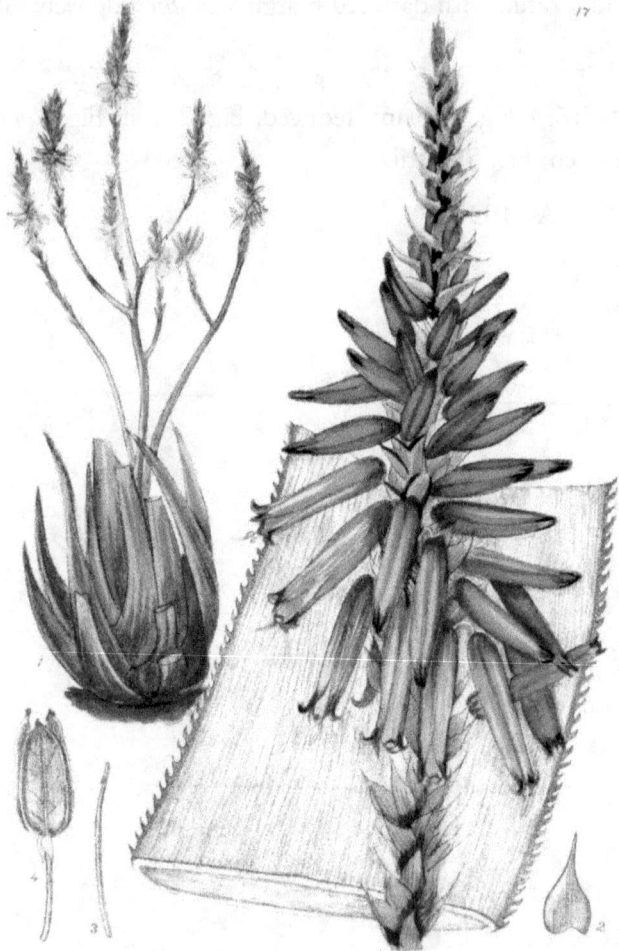

ALOE PIENAARII, POLE EVANS.

17.

K. A. Lansdell del.

ALOE PIENAARII, POLE EVANS.

PLATE 17.

ALOE PIENAARII.
Transvaal.

LILIACEAE. TRIBE ALOINEAE.

ALOE, *Linn.*; *Benth. et Hook. f. Gen. Plant.*
Aloe Pienaarii, *Pole Evans in Trans. Roy. Soc. S. Afr.*

This species was first collected by Mr. P. J. Pienaar at Smit's Drift, near Pietersburg, in January 1914, where it is very common on and around the isolated granite kopjes, though it also occurs in the open flat country. A number of plants were obtained for the gardens of the Union Buildings at Pretoria, where they have been established, and specimens are also growing in the Aloe collection at the Division of Botany Gardens, Pretoria. The species flowers from May to July.

DESCRIPTION:—*Herb*, succulent, stemless. *Leaves* 35-60 in a dense rosette, 60-80 cm. long, 12-15 cm. broad at the base, lanceolate-ensiform, acute, reddish-green or blueish, beset along the margins with small chestnut-coloured (R.C.S.) deltoid thorns 2 mm. long and 5-7 mm. apart. *Inflorescence* 2-3 from the same rosette, copiously panicled, erect, 1·25-1·65 metres high, with about 8 arcuate-erect branches subtended at the base with deltoid-acuminate bracts; racemes densely flowered, 25-35 cm. long, cylindrical-conical. *Bracts* at first densely imbricated, afterwards embracing the pedicels, 20 mm. long, 11 mm. broad, broadly ovate-acuminate, acute, many-nerved. *Pedicels* erect, spreading, 15-20 mm. long, greenish-scarlet. *Perianth* 35-38 mm.

long, somewhat 3-angled and cylindrical, at first scarlet, greenish at the tips, becoming citron-yellow (R.C.S.) when open; outer segments shorter than the inner, free, acute; inner slightly recurved at the apex and more obtuse, and the lateral ones becoming compressed towards the apex so as to close the mouth of the tube. *Stamens* just exserted; filaments bright chalcedony-yellow (R.C.S.); anthers grenadine-red (R.C.S.). *Capsule* enclosed within the dry perianth, 20 mm. long, cylindrical-trigonous, woody. *Seeds* 4-5 mm. long, irregular, narrowly winged.

PLATE 17.—Fig. 1, plant much reduced; Fig. 2, bract; Fig. 3, stamen; Fig. 4, capsule.

F.P.S.A., 1921.

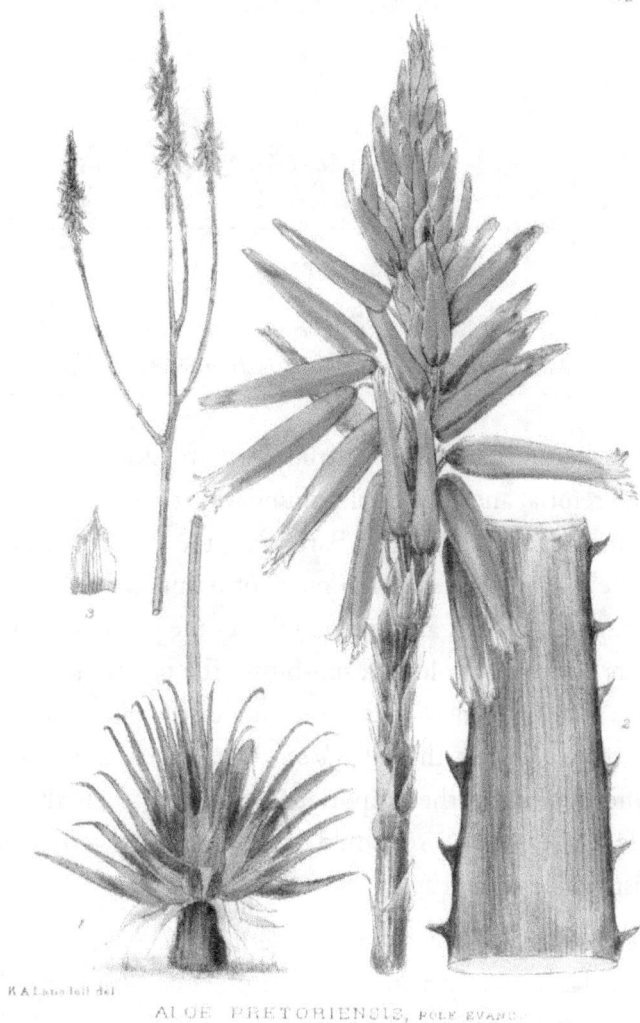

ALOE PRETORIENSIS, POLE EVANS.

18.

K. A. Lansdell del.

ALOE PRETORIENSIS, POLE EVANS.

PLATE 18.

ALOE PRETORIENSIS.
Transvaal.

LILIACEAE. TRIBE ALOINAE.

ALOE, *Linn.*; *Benth. et Hook. f.*
Aloe pretoriensis, *Pole Evans in Trans. S. Afr. Roy. Soc.*

This handsome *Aloe* occurs on the northern slopes of the hills around Pretoria, and is especially abundant on Meintjes' Kop. It is also found near Lydenburg, at Barberton, the Premier Mine, and along the foot of the Lebombo range of mountains. The flowers usually appear in May, and when in flower the plants attract large numbers of brightly coloured sun-birds. The tall branched inflorescence forms the most striking feature of the plant, and when one compares it with that of *Aloe lineata*, which is unbranched and differs in many other important respects, it seems almost incredible that *A. pretoriensis* should have been mistaken by so many botanists for *A. lineata* as has been done.

DESCRIPTION:—*Stem* short, sometimes reaching 1 metre in height, 8-12 cm. in diameter. *Leaves* numerous, 30-60 in a dense rosette, arcuate-erect, 30-65 cm. long, 3-7 cm. broad at the base, 8-10 mm. thick, lanceolate, acuminate, acute, flat on the upper surface and slightly canaliculate towards the tip, convex beneath, light green or slightly glaucous, with the margins armed with red sharply pointed horny prickles 3-4 mm. long and 10-17 mm. apart, and in old leaves the tips withered and reddish in colour. *Inflorescence* a lax panicle 2-3.5 metres high. *Peduncle* stout with 2-8 ascending branches, subtended by deltoid-ovate bracts at the base; racemes dense, 15-50 cm. long, conical-cylindric. *Bracts* at first densely imbricate, 15-20 mm. long, 10-12 mm. broad, ovate-deltoid, many veined. *Pedicels* 20-25 mm. long, lengthening and becoming erect in the fruit. *Perianth* pendulous, 40-43 mm. long, cylindrical, slightly swollen towards the middle and tapering upwards, peach-red (R.C.S.), with yellowish-green tips. *Stamens* shortly exserted; filaments greenish-yellow; anthers reddish-brown. *Style* shortly exserted. *Capsule* greyish, enwrapped in the dry perianth, 15-18 mm. long, about 6 mm. in diameter, cylindrical, 3-angled. *Seeds* dark, 2-5 mm. long, very narrowly 3-winged.

PLATE 18.—Fig. 1, plant much reduced; Fig. 2, part of a leaf, natural size; Fig. 3, bract.

F.P.S.A., 1921.

R.A.Lansdell del

CLERODENDRON TRIPHYLLUM, H.H.W.PEARSON.

19.

K. A. Lansdell del.

CLERODENDRON TRIPHYLLUM, H.H.W. PEARSON.

PLATE 19.

CLERODENDRON TRIPHYLLUM.
Transvaal, Orange Free State, Natal, Zululand.

VERBENACEAE. TRIBE VITICEAE.

CLERODENDRON, Linn.; Benth. et Hook. f. Gen. Plant.
Cyclonema triphyllum, *Harv. Thes.*

One of the charming spring plants found on the High Veld of the Transvaal and especially abundant after early winter veld fires. The corolla is of the same deep blue seen in many species of *Lobelia*, and the colour of the flowers makes the plant a conspicuous object in the veld. Our illustration was made from specimens collected by Dr. I. B. Pole Evans at Kaalfontein, between Pretoria and Germiston.

DESCRIPTION:—A low undershrub 12-60 cm. high. *Stems* erect from an underground woody rootstock, angular, striate, usually puberulous at the nodes, glabrous when mature. *Leaves* in whorls of 3 or 4, or opposite, sessile, 1·3-6 cm. long, 2-1·3 cm. broad, lanceolate or occasionally linear, acute or subacute, narrowed at the base, entire, glabrous, gland-dotted beneath. *Inflorescence* a 1-3-flowered pedunculate axillary cyme. *Peduncles* up to 2·6 cm. long, with 2 opposite lanceolate bracts near the summit. *Flowers* pedicellate. *Calyx* 3-7·5 mm. long, campanulate, 5-lobed, 5-ribbed, glabrous, with a tube equalling or slightly exceeding the ovate acute segments. *Corolla* deep chicory-blue to royal purple (R.C.S.); tube 3-7·5 mm. long, bent, villous or glabrous at the throat; 4 upper lobes unequal, obliquely obovate or elliptic, obtuse; lower lobes obovate or oblong, exceeding the upper. *Stamens* glabrous. *Fruit* a 1-2-seeded drupe, 1-1·8 cm. long, 9-1·3 cm. in diameter, ovoid, smooth.

Plate 19.—Fig. 1, fruit.

20.

K. A. Lansdell del.

GLADIOLUS REHMANNI, BKR.

PLATE 20.

GLADIOLUS REHMANNI.
Transvaal.

Iridaceae. Tribe Ixieae.

Gladiolus, *Linn.; Benth. et Hook. f. Gen. Plant.*
Gladiolus Rehmanni, *Baker; Handb. Irid. Fl.*

This species of Gladiolus is here figured for the first time. Rehmann collected it between the Elands River and Klippan, and it was then lost sight of until rediscovered by Dr. I. B. Pole Evans at Nylstroom, Waterberg District, in February, 1917, and has now been established in the Gardens of the Division of Botany, Pretoria. We are indebted to Mrs. Frank Bolus for identification.

Description:—*Corm* small, 2 cm. in diameter, subglobose, with light brown membranous tunics. *Leaves* 4-6, basal, 30-60 cm. long, 1·2 cm. broad, linear, glabrous, rigid, with prominent ribs. *Peduncle*, including the inflorescence, 60-65 cm. long. *Spike* 20-25 cm. long, lax. *Outer spathe-valve* 7-9 cm. long, oblong-lanceolate, at first bright green, then turning to dark slate-violet (R.C.S.). *Perianth* white to pale mauve (not red as stated in the *Flora Capensis*); tube curved, 2-2·5 cm. long, funnel-shaped in the upper half; lobes 4-4·5 cm. long, the 3 upper 2-2·3 cm. broad, obovate-spathulate; the 3 lower 1·8 cm. broad, oblong, with yellow-green markings at the throat. *Filaments* arcuate; anthers purple. *Style* filiform, with 3 cuneate stigmas.

PLATE 20.—Fig. 1, bulb and leaves, reduced; Fig. 2, outer spathe-valve; Fig. 3, stamens, front and side view; Fig. 4, apex of style with stigmas.

F.P.S.A., 1921.

PACHYPODIUM SUCCULENTUM.

21.

K. A. Lansdell del.

PACHYPODIUM SUCCULENTUM.

PLATE 21.

PACHYPODIUM SUCCULENTUM.
Cape Province.

APOCYNACEAE. TRIBE ECHITIDEAE.

PACHYPODIUM, *Lindl.*; *Benth. et Hook. f. Gen. Plant.*
Pachypodium succulentum, *DC., Prodr.*
Fl.Pachypodium tuberosum, Lindl., Bot. Reg. t. 1321.

The species of Pachypodium figured in our illustration was first described by the famous traveller, Carl Thunberg, in the year 1794. Thunberg gathered his plants, on which he based his description, between the Gouritz and Sundays River. The name he gave to the species was *Echites succulenta.* Robert Brown, in 1909, surmised that the plant placed by Thunberg in the genus Echites would most likely constitute a distinct genus, and in 1830 Lindley confirmed this, and founded the genus *Pachypodium* upon, and gave an excellent figure of, this species of *Pachypodium* in the *Botanical Register*, at t. 1321, but gave it a new specific name, which is omitted from the *Flora Capensis.*

Our present illustration was made from specimens growing on the rockeries of the Division of Botany, Pretoria, which were presented by Mr. Silvesta of Port Elizabeth.

DESCRIPTION:—Plants with a very large tuberous stem, partly above ground, with several semi-succulent branches arising from the upper portion of the tuber. *Branches* with a waxy covering, glabrous or finely hairy when young. *Leaves* in fascicles, 1-4 cm. long, 2-6 mm. broad, linear or linear-lanceolate, obtuse, with recurved margins, green and pubescent above, paler and tomentose below. *Spines* arising in groups of 2-3 from an evident cushion, the two lateral spines longer and spreading, the medium spine shorter and erect, sometimes absent. *Flowers* terminal. *Calyx* campanulate; lobes narrowly lanceolate, acute, densely pubescent. *Corolla* twisted in bud; 1-1·5 cm. long, cylindric, pubescent; lobes 1·5 cm. long, 1 cm. broad, obovate, narrowed into a distinct claw, pale pink with dark-red markings. Fruit 6-8 cm. long, spindle-shaped, pubenulous.

PLATE 21.—Fig. 1, plant much reduced; Fig. 2, calyx; Fig. 3, stamen; Fig. 4, style and stigma.

F.P.S.A., 1921.

PROTEA ABYSSINICA

22.

K. A. Lansdell del.

PROTEA ABYSSINICA.

PLATE 22.

PROTEACEAE. Tribe PROTEEAE.

Protea, *Linn.*; *Benth. et Hook. f. Gen. Plant.*

Protea abyssinica, *Willd. Sp. Pl. vol. i. p. 522; Fl.*

The Protea illustrated here is a very common species on the hill-sides in the neighbourhood of Pretoria. It sometimes attains a height of 15 feet, is much branched, and has no distinct trunk. We have no record of the species occurring further south, but it certainly extends into Rhodesia, and perhaps—though we have some doubt on this point—into Abyssinia. The species was first described by the botanist Willdenow, under the present specific name in 1797, and he based his description on a figure which appeared in Bruce's *Travels to discover the Source of the Nile*, which was published in 1790. The point as to whether the Transvaal plant is the same species as that figured by Bruce needs further investigation.

The specimens from which the figure was made were collected by Miss I. C. Verdoorn at Waterkloof, near Pretoria.

DESCRIPTION:—*Branches* glabrous. *Leaves* 7-15·5 cm. long, ·8-2·2 cm. broad, narrowly oblong-lanceolate or lanceolate, subacute or obtuse, narrowing to the base, coriaceous, glabrous. *Inflorescence,* 6·3 cm. long, and about 6·3 cm. in diameter when expanded, narrowed into a short scaly stipe. *Involucral-bracts*

11-seriate, silky; the inner oblong, concave, shorter than the flowers. *Perianth* with three small teeth at the apex, densely hairy. *Ovary* covered with a dense tuft of long hairs; style 4·5 cm. long, more or less curved; stigma slightly bent at the junction with the style.

F.P.S.A., 1921.

23.

K. A. Lansdell del.

BOLUSANTHUS SPECIOSUS, HARMS.

PLATE 23.

BOLUSANTHUS SPECIOSUS.
Transvaal, Rhodesia, Portuguese East Africa.

LEGUMINOSAE. TRIBE SOPHOREAE.

BOLUSANTHUS, *Harms in Fedde Repert. Nov. Sp.*
Bolusanthus speciosus, *Harms. l.c.*
Lonchocarpus speciosus, Bolus in Journ. Linn. Soc.

This remarkable and handsome leguminous plant was collected by the late Dr. Bolus near the Komati River Drift in 1886, and described by him as *Lonchocarpus speciosus.* Dr. Harms of Berlin, when examining a collection of Rhodesian plants, came across the same species on which he founded the genus *Bolusanthus.* The free stamens would indicate that it is not a species of *Lonchocarpus.*

Our illustration was made from material collected by Dr. Pole Evans at Chunies Poort, Transvaal, in October, 1919. The tree, which frequently reaches a height of 30-40 feet, is locally known as "Van Wyk's Hout," or "Wild Wisteria." It is frequent along the northern foothills of the Zoutpansberg range of mountains and in the low veld bush country along the Selati River. When in full bloom it is one of the most beautiful sights seen in the veld, and is a species which should certainly be introduced into cultivation.

DESCRIPTION:—Tree up to 30-40 ft. high. *Branchlets* pubescent. *Leaves* 10-27 cm. long; leaflets petiolate; petiole 5 mm. long; leaflet 2·5-7·5 cm. long, ·5-2·5 cm. broad, ovate-elliptic or lanceolate, very often sub-falcate, long-acuminate, acute, oblique at the base, villous when young, becoming pubescent with age. *Inflorescence* a raceme 14-20 cm. long; rachis pubescent. *Pedicels* up to 2 cm. long, pubescent. *Calyx* 7 mm. long, densely tomentose. *Corolla* dark blue; vexillum 1·5 cm. long, about 1·3 cm. broad, obovate; alae 1·3 cm. long, carina as long as the alae. *Stamens* free. *Ovary* linear, densely pubescent. *Fruit* up to 7 cm. long, 1-1·2 cm. broad, linear-oblong.

PLATE 23.—Fig. 1, branch with flowers; Fig. 2, leaf; Fig. 3, legumes; Fig. 4, calyx; Fig. 5, vexillum; Fig. 6, alae; Fig. 7, carina; Fig. 8, ovary.

F.P.S.A., 1921.

ACOKANTHERA SPECTABILIS, HOOK. F.

24.

K. A. Lansdell del.

ACOKANTHERA SPECTABILIS, HOOK. F.

PLATE 24.

ACOKANTHERA SPECTABILIS.
Cape Province, Natal.

APOCYNACEAE. TRIBE CARISSEAE.

ACOKANTHERA, *G. Don.*; *Benth. et Hook. f. Gen. Plant.*
cokanthera spectabilis, *Hook. f. Bot. Mag. t. 6359; Fl.*

The above figure in the *Botanical Magazine* was published in 1878, and together with the description which accompanied it was the first recognition that the so-called "Gift Boom" of the Eastern Province consisted of two distinct species. The plant from which the figure in the *Botanical Magazine* was made, flowered at Kew Gardens in 1878. Mr. T. R. Sim states that he cannot distinguish *A. spectabilis*, Hook. from *A. venenata*, G. Don., but regards it as an eastern coastal form. The plant is reputed to be extremely poisonous, and as the fruits are so attractive-looking, it makes the species also a dangerous one. In habit our plant is an evergreen shrub which lends itself to cultivation in the shrubbery; the flowers are very fragrant, and even in fruit the shrub does not lose its beauty, as the dark purple fruits show up conspicuously against the green leaves. The specimen figured here was presented by Mr. J. W. Wickens from the Garden of the Union Buildings, Pretoria.

DESCRIPTION:—Shrub 4-10 ft. high. *Branches* glabrous. *Leaves* shortly petioled; petioles 6 mm. long; lamina 7-10 cm. long, 2·2-4 cm. broad, elliptic or elliptic-lanceolate, shortly acuminate, acute, narrowed at the base, glabrous, with the mid-rib distinct below and sunken above. *Flowers* in many-flowered clusters. *Calyx* 5-lobed almost to the base, pubescent. *Corolla-tube* 2 cm. long, narrowly cylindric, pubescent outside, hairy within at the throat; lobes 4 mm. long, 3 mm. broad, elliptic, rounded at the apex. Anthers ovate in outline, with a few hairs at the apex. Style 1·4 mm. long, cylindric; stigma subglobose, with a few hairs at the apex.

[*A. spectabilis* is very easily distinguished from *A. venenata* by its longer petioles, usually larger size, and less elliptic shape and different venation of its leaves, the veins (at least in the dried state) being far less prominent and less ascending than they are in *A. venenata*, and the flowers are much larger, the corolla-tube of *A. spectabilis* being 14-20 mm. long, whilst those of *A. venenata* are only 8-12 mm. long. Dried specimens show no intermediates.—N. E. BROWN.]

PLATE 24.—Fig. 1, calyx; Fig. 2, corolla in section; Fig. 3, stamen; Fig. 4, stigma.

F.P.S.A., 1921.

CYRTANTHUS SANGUINEUS, HOOK.

25.

K. A. Lansdell del.

CYRTANTHUS SANGUINEUS, HOOK.

PLATE 25.

CYRTANTHUS SANGUINEUS.
Cape Province, Transkei, Natal.

AMARYLLIDEAE. TRIBE AMARYLLEAE.

CYRTANTHUS, Ait.; Benth. et Hook. f. Gen. Plant.
Cyrtanthus sanguineus, *Hook. in Bot. Mag. t. 5218; Fl.*

This species was imported into England from Kaffraria by Messrs. Backhouse, and presented by them to the Horticultural Society of London in 1846. Two years later, in the *Journal* of the Society, Dr. Lindley described the plant, and the description was accompanied by a woodcut. In 1860 a plant flowered in the greenhouse at Kew, and was figured and described in the *Botanical Magazine*, t. 5218, by Hooker. The specimens from which the present illustration was made were gathered by Miss K. A. Lansdell at Krantzkloof, Natal. The plant is known as the "Kei Lily."

DESCRIPTION:—Bulb about 4·5 cm. long, ovoid, produced into a distinct neck, with parchment-like scales. Leaves 1-4, contemporary with the flowers, 22-30 cm. long, ·5-2 cm. broad, linear, obtuse, narrowed more or less suddenly in the lowermost third to form a petiole, glabrous. *Peduncle* 12-26 cm. long, bearing 1-2 flowers. *Spathes* two, 4-6·5 cm. long, tapering into a long appendage from a broadened base. *Pedicels* up to 2 cm. long, glabrous. *Perianth-tube* 4-6 cm. long, campanulate in the upper half, cylindric in the lower half; lobes 4 cm. long, 1·8 cm. broad, ovate, acuminate, acute, with the apices of the outer lobes inflexed and forming a small hood. *Stigmas* with minute papillae on their inner faces.

PLATE 25.—Fig. 1, portion of apex of perianth-lobe; Fig. 2, upper portion of style showing stigmas.

F.S.P.A., 1921.

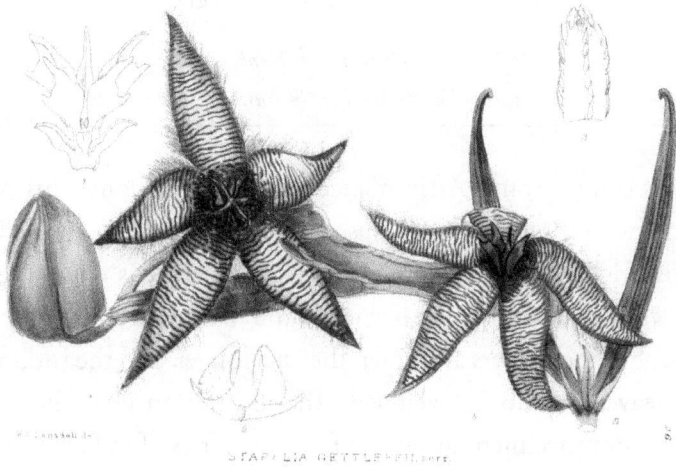

STAPELIA GETTLEFFII, Pott.

26.

K. A. Lansdell del.

STAPELIA GETTLEFFII, POTT.

PLATE 26.

STAPELIA GETTLEFFII.
Transvaal.

ASCLEPIADACEAE. TRIBE STAPELIEAE.

STAPELIA, Linn.; Benth. et Hook. f. Gen. Plant.
Stapelia Gettleffii, *Pott in Ann. Transvaal Mus.*

Lovers of our South African succulents will welcome this plate of a new Transvaal *Stapelia*, discovered by Mr. G. F. Gettleffi at Louis Trichardt in the Zoutpansberg District. It is closely allied to *Stapelia hirsuta*, which occurs in the Western Province of the Cape, but the flowers are larger, the cilia longer, and the rudimentary leaves are more developed. The illustration given here was made from specimens growing on the rockeries of the Division of Botany, Pretoria, but there is no record of the locality from which the original plants came. In 1916 a coloured illustration of the species appeared in the *Botanical Magazine* (t. 8681), made from a specimen which flowered in the Royal Botanic Gardens, Kew, in June 1915, which was sent to England by Mr. N. S. Pillans. Mr. Pillans' specimens came from Palapye Road, near Mafeking.

DESCRIPTION:—A succulent herb 10-20 cm. high. *Stems* decumbent, 4-angled, velvety-pubescent. *Leaves* rudimentary, ·3-1·3 cm. long, linear-lanceolate, acute, velvety-pubescent. *Flowers* 1-3 together near the base of the stem; pedicels velvety. *Sepals* velvety. *Corolla* 8·5-15 cm. in diameter; disc purple, clothed with long soft hairs; lobes barred with transverse yellow and purple lines, and ciliate with long whitish and purple hairs, velvety on the back. Outer corona-lobes 7 mm. long, lanceolate with a subulate-acuminate recurved dark purple tip; inner corona-lobes ·9-1·3 cm. long, subulate, with a 1-3-toothed broad dorsal wing.

[As received from South Africa and as grown in England the stems of all the plants seen are erect, being decumbent only at the basal part as in other species of this genus. I have never seen them entirely prostrate as here represented. Locality may cause the difference.—N. E. BROWN.]

PLATE 26.—Fig. 1, corona; Fig. 2, pollinia; Fig. 3, upper portion of stem; Fig. 4, stem with flowers; Fig. 5, follicles.

F.P.S.A., 1921.

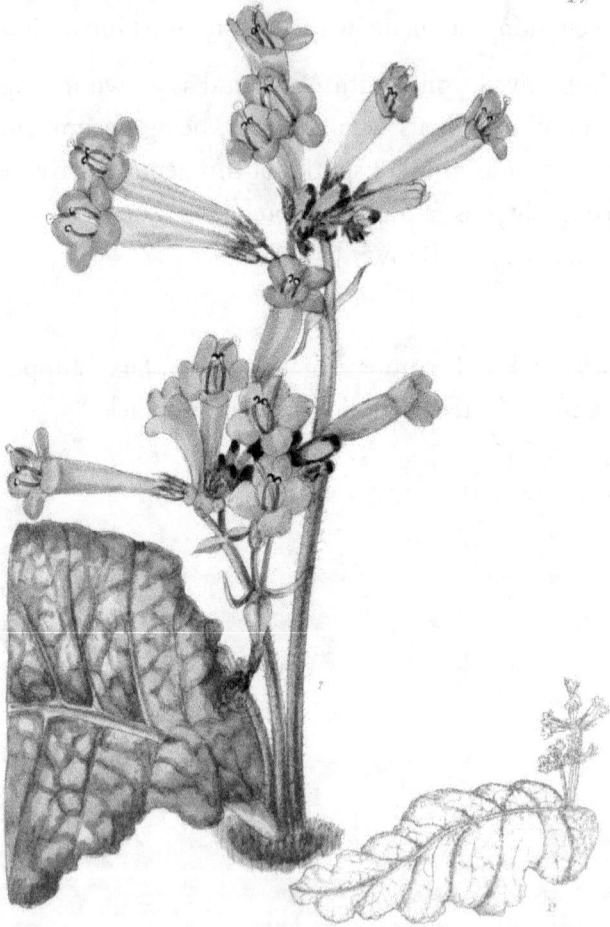

27.

K. A. Lansdell del.

STREPTOCARPUS DUNNII, HOOK. f.

PLATE 27.

STREPTOCARPUS DUNNII.

Transvaal.

GESNERACEAE. TRIBE CYRTANDREAE.

STREPTOCARPUS, Lindl.; Benth. et Hook. Gen. Plant.
Streptocarpus Dunnii, *Hook. f. Bot. Mag. t. 6903; Fl.*

This species of *Streptocarpus*, which belongs to a section of the genus characterised by the development of one leaf only, was first brought to the notice of horticulturists in 1884 by Mr. E. G. Dunn, who sent seeds to Kew from Spitzkop in the Transvaal.

The seeds germinated freely, and in May and June of 1886 the plants were a feature of the Succulent House at Kew. The genus *Streptocarpus* is well represented in South Africa, and at least 24 distinct species are known.

Our illustration was made from plants grown by Mr. C. E. Gray, Pretoria, from specimens collected by Dr. Pole Evans on Mr. Geo. Heys' farm, Weltevreden, Machadodorp, where it grows on rocks at the side of a stream.

DESCRIPTION:—Leaf sometimes up to 1 m. long and 45 cm. broad, hairy beneath, sometimes shaggy on the upper surface, with crenate margins. *Peduncles* up to 15 cm. long, terete, pilose, bearing many flowers arranged more or less in a cymose manner. *Calyx* divided almost to the base; lobes 5 mm. long, lanceolate or linear-lanceolate, subacute, ciliate. *Corolla-tube* 2·2 cm. long, pubescent in bud, becoming more or less glabrous with age, gradually widening from the base upwards; lobes 4 mm. long, 4 mm. broad, more or less transversely oblong, broadly rounded at the apex. *Style* densely pilose below.

PLATE 27.—Fig. 1, inflorescence; Fig. 2, plant reduced

F.P.S.A., 1921.

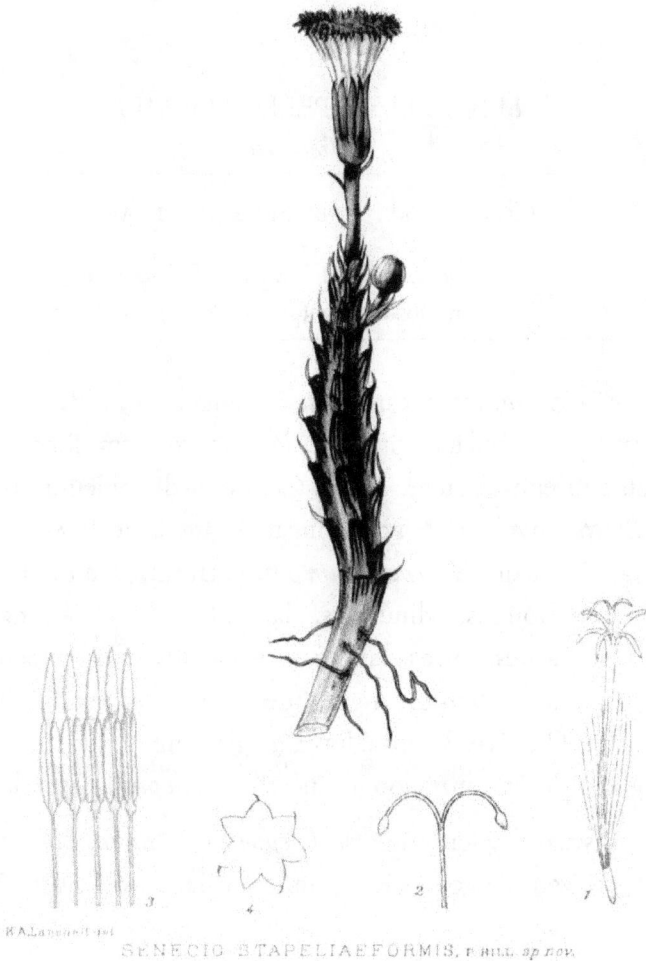

SENECIO STAPELIAEFORMIS, P. HILL. *sp. nov.*

28.

K. A. Lansdell del.

SENECIO STAPELIAEFORMIS, P. HILL. *sp. nov.*

PLATE 28.

SENECIO STAPELIAEFORMIS.
Transvaal.

COMPOSITAE. TRIBE SENECIONIDEAE.

SENECIO, Linn.; Benth. et Hook. f. Gen. Plant.
Senecio stapeliaeformis, *Phill. sp. nov.*

Caudex 7-25 cm. altus, carnosus, 4-7-angulosus. *Folia* 2-5 mm. longa, erecta, subulata, emarcida. *Pedunculus* 2 cm. longus, simplex, monocephalus, teres, glaber. *Capitulum* discoideum, coccineum. *Bractae* involucri, 1·5 cm. longae, 1 mm. latae, lineares, apice obtusae ciliataeque. *Receptaculum* 3 mm. latum, planum. *Corollae* tubus 2 cm. longus, cylindricus, glaber; lobi 3·5 mm. longi, ·75 mm. lati, lineares, apice obtusi. *Stamina* inclusa; filamenta 6 mm. longa; antherae 2·5 mm. longae, lineares, apice appendice lineare instructae. *Ovarium* 2·5 mm. longum, glabrum; stylus 2 mm. longus, glaber, lobis 4 mm. longis linearibus. *Pappus* 1·6 cm. longus.

Transvaal: Lydenburg. *Carl Jeppe in National Herbarium,* 1272. Pruizen, Potgeiters Rust, under bushes, *Burtt Davy,* 2203.

The specimens from which our figure was made were collected by Mr. Carl Jeppe in the Lydenburg District, Transvaal, and flowered in the Garden of the Division of Botany, Pretoria, in September, 1919.

It is closely allied to *Senecio pendula,* Sch. Bip., a native of Somaliland and Arabia, but differs in the erect, angled stems.

The stems resemble those of a *Stapelia* to such an extent that it was thought to be a *Stapelia* when received, and was planted out in the *Stapelia* collection. This species will make a very welcome addition to the South African rockeries.

DESCRIPTION:—*Stems* 7-25 cm. long, simple, more rarely branched, thick and fleshy, 4-7-angled, with the angles compressed and toothed, each tooth tipped with an erect, slender, awl-like leaf 2-5 mm. long, withering and becoming hardened. *Peduncle* often solitary and terminal, sometimes there is also an axilliary one on the stem, but then only the uppermost appears to develop; 2 cm. long, bearing one flower-head, terete, with 3-4 of the subulate leaves or bracts, glabrous. *Flower-head* solitary, discoid, scarlet-red. *Involucral-scales* in a single row of 10-12, more or less concrete, 1·5 cm. long, 1 mm. broad, linear, obtuse, ciliated at the apex, brick-red. *Receptacle* 3 mm. in diameter, flat. *Florets* all hermaphrodite. *Corolla-tube* 2 cm. long, cylindric, glabrous, scarlet-red above, colourless below; lobes 3·5 mm. long, 0·75 mm. broad, linear, obtuse, scarlet-red. *Stamens* inserted about halfway down the corolla-tube; filaments 6 mm. long, filiform, becoming linear for 1·5 mm. below the anthers; anthers 2·5 mm. long, linear, blunt at the base, tipped at the apex with a linear appendage 1·5 mm. long. *Ovary* 2·5 mm. long, linear in outline, glabrous; style 2 mm. long, cylindric, glabrous; style-arms 4 mm. long, linear, tipped with a bristly cone. *Pappus* of white hairs 1·6 cm. long, distantly barbellate. *Fruit* not seen.

[A few years before the war a plant of this species was sent by Mr. J. Burtt Davy to Kew Gardens, where it flowered annually, but has since died.—N. E. BROWN.]

I. B. Pole Evans

PLATE 28.—Fig. 1, flower (enlarged); Fig. 2, style-arms; Fig. 3, stamens; Fig. 4, cross-section of stem.

F.P.S.A., 1921.

NYMPHAEA STELLATA with.

29.

K. A. Lansdell del.

NYMPHAEA STELLATA WITH.

PLATE 29.

NYMPHAEA STELLATA.
Cape Province, Transvaal, Natal, Rhodesia.

NYMPHAEACEAE. TRIBE NYMPHAEAE.

NYMPHAEA, Linn.; Benth. et Hook. f. Gen. Plant.
Nymphaea stellata, *Willd. Sp. Pl. vol. ii. p. 1153; Fl.*

A common water plant in many of our South African rivers and vleis, and it is not surprising that such a handsome species soon found its way to cultivators in Europe. Masson, about the year 1792, appears to have first introduced it into England by forwarding specimens from the Cape to the Royal Gardens at Kew. It was not long before coloured plates appeared in the botanical publications of the day, and the first of these was published in 1801 in the *Botanical Magazine* and about the same time in Andrews' *Botanist's Repository*. A second figure again appeared in the *Botanical Magazine* about 18 years later. The species, commonly known as the "Blue Water Lily" (Zulu "i-Ziba"), is easy of cultivation, and is found in most garden ponds in South Africa. Our illustration was made from specimens growing in the aquarium of the Natal Herbarium, Durban.

DESCRIPTION:—An aquatic plant with a submerged rhizome from which the floating leaves and flowers are produced. *Rhizome* 4-5 cm. in diameter, black and spongy. *Leaves* about 6 to each rhizome; petiole long or short according to the depth of the water, terete, striate, thickly clothed with transparent hairs; lamina green above, brownish beneath, up to 30 cm. long and 20-26 cm.

broad, orbicular or elliptic, rounded at the apex, and with a deep acute triangular notch at the base, with entire or sometimes wavy margins, and prominent veins beneath, glabrous. *Peduncles* longer than the petioles, raising the flower well above the surface of the water. *Sepals* 4, green outside, blue within, 4-6 cm. long, 1·5-2 cm. broad, ovate-oblong, acuminate. *Petals* numerous, about 4 cm. long, 1 cm. broad, ovate-lanceolate, obtuse, blue. *Torus* thick, fleshy. *Stamens* numerous, in several rows; filaments flattened; the outer longer than the inner, and ½-⅔ the length of the petals; anthers yellow, with a long linear blue appendage at the apex. *Carpels* many, inserted in the torus; stigma arcuate, obtuse. *Fruit* a many-seeded berry. *Seeds* spongy.

PLATE 29.—Fig. 1, torus; Fig. 2, plant reduced.

F.P.S.A., 1921.

30.

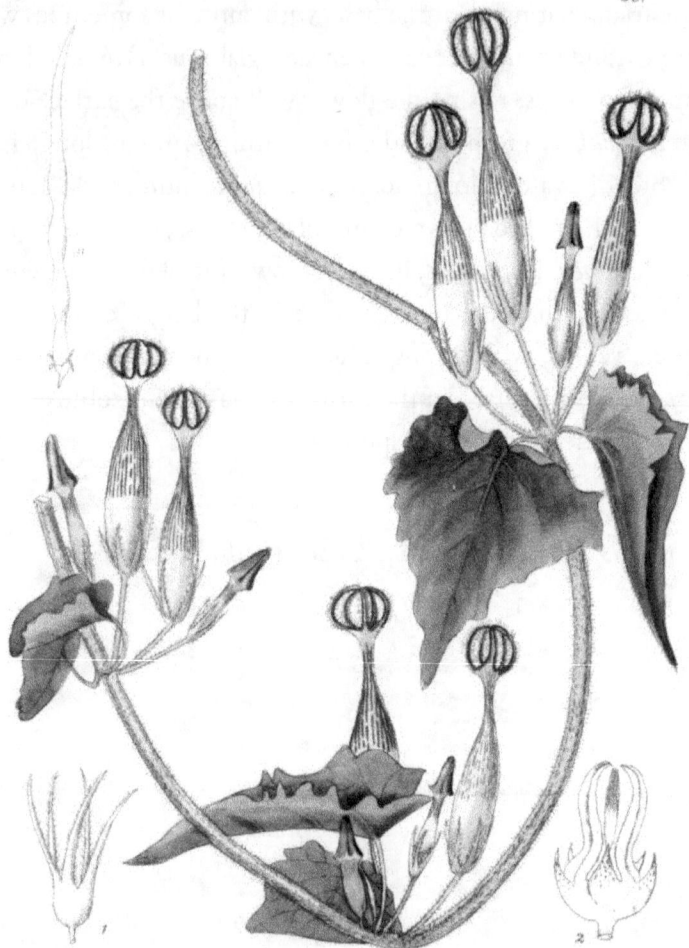

CEROPEGIA MEYERI DENE.

K.A.Lansdell del.

30.

K. A. Lansdell del.

CEROPEGIA MEYERI DENE.

PLATE 30.

CEROPEGIA MEYERI.

Cape Province, Transvaal, Natal.

ASCLEPIADACEAE. TRIBE CEROPEGIEAE.

CEROPEGIA, *Linn.; Benth. et Hook. f. Gen. Plant.*

Ceropegia Meyeri, *Decne. in DC. Prodr.*

This species was first collected by Drège between the Bashee River and Morley, in Tembuland, about the year 1831, but the species has been found by several collectors since then. Its altitudinal range of distribution is wide, as it has been recorded by the late Dr. Wood from the sub-tropical climate of Durban and from Oliver's Hoek Pass on the Drakensbergen, which is occasionally covered with snow in the winter months. The plant is a very ornamental twiner, easily cultivated, and well worth the attention of horticulturists. The accompanying illustration was made from specimens growing in the garden of the Natal Herbarium at Durban.

DESCRIPTION:—*Rootstock* a flattened tuber. *Stem* herbaceous, twining, up to 10 metres long, pubescent. *Leaves* petioled; lamina 1·7-2·8 cm. long, 3-5·9 cm. broad, cordate-ovate or lanceolate-ovate, acute, somewhat acuminate, cordate or rounded at the base, more or less pubescent or rarely subglabrous on both sides, with the margins variously toothed or lobed. *Petiole* 1-4 cm. long, pubescent. *Inflorescence* 2-4-flowered, cymose, sessile or subsessile at the nodes. *Pedicels* 0·6-1·1 cm. long, villous. *Sepals* 8-11 mm. long, 1 mm. broad at the base, subulate, pubescent. *Corolla-tube* whitish at the lowermost third, streaked with purple lines above,

2·5-3·1 cm. long, bottle-shaped, inflated and cylindric-oblong in the basal two-thirds, and narrowed into a cylindric neck above, then abruptly dilated at the mouth, glabrous without and within except at the throat; lobes almost black, connate at the tips, 1-1·2 cm. long, 3 mm. broad at the base, linear, pilose, with reflexed margins. *Outer corona-lobes* white, ascending, 1 mm. long, deltoid, acute, glabrous; inner corona lobes connivent at the base, then slightly divergent, and again connivent at the tips, white above, black below, 2 mm. long, linear or slightly spathulate-linear, obtuse. *Follicles* erect, sub-parallel, 8-10 cm. long, tapering into a beak, glabrous.

PLATE 30.—Fig. 1, calyx (enlarged); Fig. 2, corona; Fig. 3, a follicle.

F.P.S.A., 1921.

31.

K. A. Lansdell del.

MORÆA IRIDIOIDES, LINN.

PLATE 31.

IRIDEAE. TRIBE MORAEEAE.

MORAEA, Linn.; Benth. et Hook. f. Gen. Plant.
Moraea iridioides, *Linn. Mant. 28; Fl.*

This is one of the largest and most handsome species in the genus and is frequently cultivated in South African gardens. Thunberg appears to have been the first collector of this plant; he gathered his specimens near the Zeekoe River in Humansdorp Division about 1772, but the species was known in England before then, as there is a record of Miller having it in cultivation in 1758. The first figure of the species appeared in the *Botanical Magazine* in 1804 and it has been figured several times since. The present illustration was made from specimens growing in the garden of the Natal Herbarium, Durban.

DESCRIPTION:—A perennial plant with short underground rhizomes. *Leaves* crowded in dense fan-shaped basal rosettes, 0·6-1·3 metres long, 1-2 cm. broad, linear, acute, equitant at the base, glabrous. *Peduncles* equalling or exceeding the leaves. *Inflorescence* corymbose. *Spathe-valves* 2, about 6·5 cm. long, obtuse, tightly folded; the outer smaller than the inner. *Perianth-segments* 5-6 cm. long, 2-3 cm. broad, obovate, obtuse, clawed at the base; the 3 outer segments with an orange-yellow keel, densely pilose at the base; the 3 inner segments narrower, with dark markings above the claw. *Ovary* ellipsoid. *Stigmas* purple, lanceolate, 2-lobed. *Fruit*

5 cm. long, 1·7 cm. in diameter, ellipsoid; valves coriaceous. *Seeds* discoid.

F.P.S.A., 1921.

HAEMANTHUS NATALENSIS, PAPPE

32.

K. A. Lansdell del.

HAEMANTHUS NATALENSIS, PAPPE.

PLATE 32.

HAEMANTHUS NATALENSIS.
Cape Province, Natal.

AMARYLLIDACEAE. TRIBE AMARYLLEAE.

HAEMANTHUS, Linn.; Benth. et Hook. f. Gen. Plant.
Haemanthus natalensis, *Pappe ex Hook. in Bot. Mag. t. 5378; Fl.*

The late Dr. Pappe first brought this species to the notice of Kew as an undescribed South African plant, and not long afterwards (1862), Dr. Sanderson sent bulbs from Natal to the Royal Botanic Gardens, Kew, which flowered the following year. An excellent figure of the plant appeared in the *Botanical Magazine* of the same year. The species appears to be fairly common in Natal, but the only Cape Province record we have is supplied by a specimen collected by Mr. W. Tyson at Kokstad, East Griqualand, 1883. The present illustration was made from specimens collected by Miss K. A. Lansdell at "Stella Bush" near Durban. It is popularly known as the "Blood Flower," "Snake Lily," and "April Fool." It is reputed to be poisonous, but is used medicinally by the natives of Natal who know it as "Indumbe-ka-Hloile."

DESCRIPTION:—An erect plant about 1 metre high. *Bulb* 2-7.5 cm. in diameter, usually globose. *Stem* about 1 metre high, closely covered with leaves above and with a few scale-leaves at the base. *Leaves* sub-erect, 32 cm. long, 8-9 cm. broad, acute, narrowed at the base, glabrous, shining; the sheathing petiole of the lowermost leaves with reddish-brown spots, and the margin round the apex coloured reddish-brown. *Peduncle* lateral, from the base of the stem, generally shorter than the stem, semi-terete, smooth and glabrous. *Inflorescence* a many-flowered umbel. *Involucral-bracts* 7-8, vandyke red to blackish red-purple (R.C.S.), 6·5-7·5 cm. long, 2-5 cm. broad, oblong, sub-acuminate, rounded or obtuse at the apex, glabrous. *Floral-bracts* about 4 cm. long, 1-2 mm. broad, linear. *Flowers* scarlet (R.C.S.). *Pedicels* 1·5-5 cm. long, terete, glabrous. *Perianth-tube* 1 cm. long, 0·9 cm. in diameter, campanulate, glabrous; lobes 1·2 cm. long, linear, obtuse and recurved at the apex, with a tuft of hairs on each alternate lobe, otherwise glabrous. *Stamens* exserted, arising at the throat of the perianth-tube; filaments usually about 1·6 cm. long, hermosa pink (R.C.S.); anthers yellow. *Ovary* 6 mm. long, ellipsoid; style longer than the stamens, hermosa pink (R.C.S.); stigma minute, globose. *Fruit* a bright-red berry about 1 cm. in diameter, sub-globose, 1-3 seeded.

PLATE 32.—Fig. 1, plant reduced; Fig. 2, flower, with bract; Fig. 3, stamen, showing attachment to segment of the perianth.

F.P.S.A., 1921.

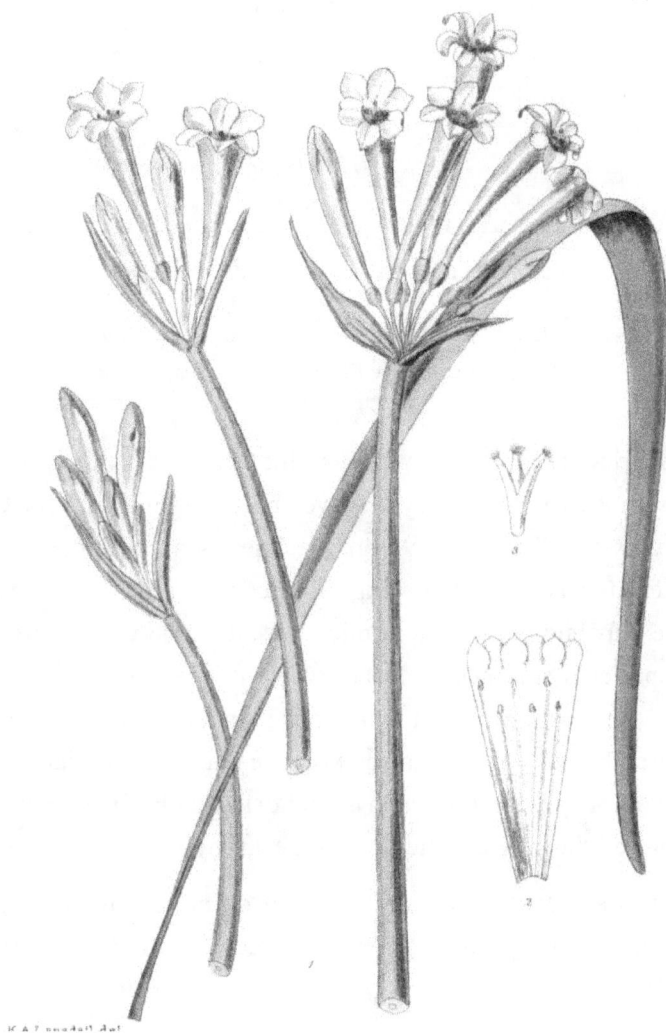

33.

K. A. Lansdell del.

CYRTANTHUS MᶜKENII, Hook. f.

PLATE 33.

CYRTANTHUS McKENII.
Natal.

AMARYLLIDACEAE. TRIBE AMARYLLEAE.

CYRTANTHUS, *Ait.; Benth. et Hook. f. Gen. Plant.*
Cyrtanthus McKenii, *Hook. fil. in Gard. Chron.*

The species of *Cyrtanthus*, which with one exception are confined to South Africa, have always received notice from gardeners. Our plant was described in 1869 from specimens sent to Europe by Mr. McKen, and a coloured drawing appeared soon after (1873) in one of the illustrated botanical publications. As far as our records go, this species is confined to Natal, where it is known as the "Ifafa Lily." The specimens from which our illustration was made were gathered by Miss K. A. Lansdell on the banks of the Ifafa River near Port Shepstone, the original locality in which Mr. McKen first discovered the species. The flowers are strongly scented.

DESCRIPTION:—*Bulb* 3-4 cm. in diameter, ovoid; tunics brown, membranous. *Leaves* 2-6, erect, contemporary with the flowers, 20-30 cm. long, 0·9-2 cm. broad, linear, obtuse, narrowed to the base, glabrous. *Peduncle* reddish-brown near the base, longer than the leaves, sub-terete, hollow. *Inflorescence* a 4-10-flowered umbel. *Spathe-valves* 2, green, spotted with reddish-brown marks when young, at length withering, 2·5-3·5 cm. long, 4-7 mm. broad, lanceolate, acute. *Flowers* sub-erect, pure white with yellowish throats. *Floral-bracts* linear. *Pedicle* 0·8-1·5 cm. long, terete. *Perianth-tube* 3-3·5 cm. long, 7-9 mm. in diameter

102

at the throat, gradually widening from the base upwards; lobes spreading, 6-7 mm. long, 5-6 mm. broad, ovate, the 3 outer cucullate at the apex; the 3 inner emarginate. *Stamens* sub-sessile, in 2 whorls below the throat of the perianth tube; anthers oblong. *Ovary* sub-trigonous; style exserted; stigmas spreading, oblong-linear, tufted at the apex. *Fruit* a trigonous capsule.

PLATE 33.—Fig. 1, leaf and flowers, natural size; Fig. 2, perianth laid open; Fig. 3, apex of style showing stigmas.

F.P.S.A., 1921.

WITSENIA MAURA, Thunb.

34.

S. Gower del.

WITSENIA MAURA, Thunb.

PLATE 34.

WITSENIA MAURA.
Cape Province.

IRIDACEAE. TRIBE SISYRINCHIEAE.

WITSENIA, Thunb.; Benth. et Hook. f. Gen. Plant.
Witsenia maura, *Thunb. Nov. Gen. pl. p. 34; Fl.*

This interesting plant, the only species known in the genus, was first found by Dr. Carl Thunberg at Noordhoek and False Bay on the Cape Peninsula and described by him in 1782. It appears to be confined to damp habitats in the Cape Province, and has been found by the late Dr. Bolus at Houw Hoek in Caledon Division. It has also been recorded from the Tradouw Mountains in Swellendam Division, and this year (1920) Mr. T. P. Stokoe has discovered the plant on the Klein River Mountains near Caledon. It is a rare species and would only interest enthusiastic cultivators on account of its rarity.

We are indebted to Mr. Stokoe for the living specimens from which this plate was prepared. The plant is known locally as "Waaiertje."

DESCRIPTION:—*Stems* woody. *Leaves* distichous, about 18 cm. long, 5-7 mm. broad, linear, tapering to an acute point, amplexi-caul, glabrous. *Flowers* in terminal heads. *Bracts* 4·5 cm. long, boat-shaped, shorter than the flowers. *Perianth-tube* 2·7 cm. long, brown below, becoming blue-black above; lobes 1·4 cm. long, 5·5 mm. broad, ovate-lanceolate, obtuse, densely tomentose with yellow hairs outside, glabrous within, with a tuft of yellow hairs at the apex of the inner segments and marginal hairs round the apex of the outer segments. *Stamens* inserted near the throat of the perianth-tube; filaments 5 mm. long, linear and slightly dilated at the base; anthers 6 mm. long, linear. *Ovary* small; style 4 cm. long, slightly bifid at the apex.

PLATE 34.—Fig. 1, plant natural size; Fig. 2, unopened flower; Fig. 3, lobes of perianth; Figs. 4 and 5, stamens; Fig. 6, ovary and style; Fig. 7, tip of style, showing the three stigmas.

F.P.S.A., 1921.

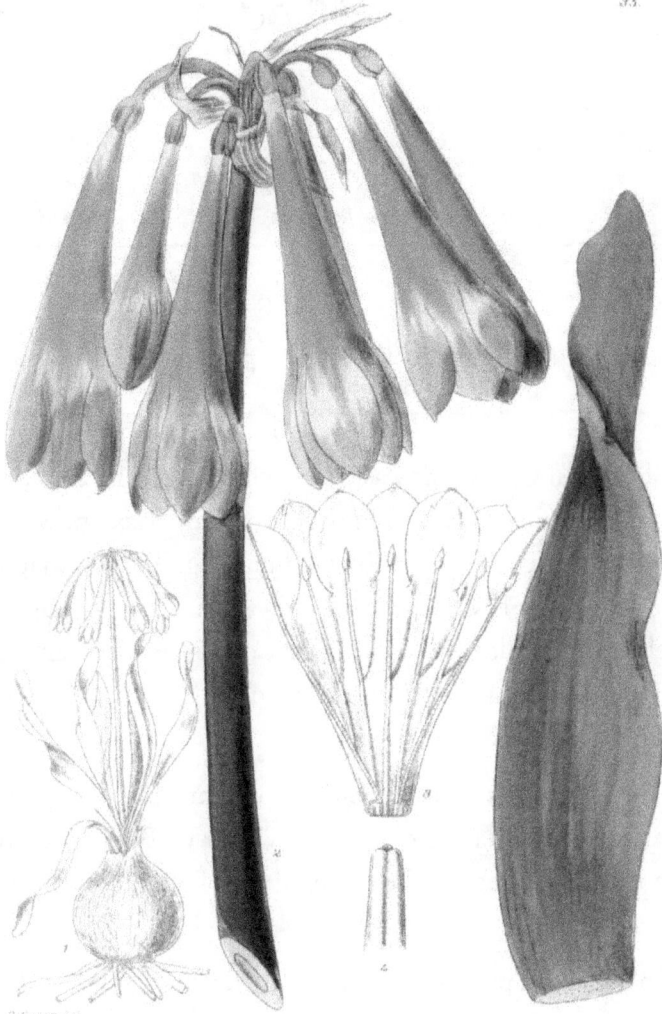

CYRTANTHUS OBLIQUUS, Ait.

35.

S. Gower del.

CYRTANTHUS OBLIQUUS, Ait.

PLATE 35.

CYRTANTHUS OBLIQUUS.
Cape Province, Natal.

AMARYLLIDACEAE. TRIBE AMARYLLEAE.

CYRTANTHUS, *Ait.; Benth. et Hook. f. Gen. Plant.*
Cyrtanthus obliquus, *Ait.; Hort. Kew.*

This beautiful *Cyrtanthus* was described by Aiton in 1789, probably from plants collected in South Africa by Masson, who sent specimens of this species to England in 1774. The fact that it has been so frequently figured is an indication that it has appealed largely to horticulturists. Jacquin first produced a coloured plate in 1797 and the last figure we know of in botanical literature is that quoted above.

The specimens from which the present plate was prepared flowered in the Gardens of the Division of Botany, Pretoria, from bulbs gathered on the mountains at Bethelsdorp near Port Elizabeth by Dr. I. B. Pole Evans.

In Natal this plant is known as "Justifina" or "Sore-eye flower" by the natives, who use it medicinally, as "Matoonga."

DESCRIPTION:—*Bulb* globose, about 10 cm. in diameter, with a short neck about 4 cm. long, and thick cylindrical roots from the base; outer tunics membranous. *Leaves* 18 cm. long, 4 cm. broad, strap-shaped, obtuse, glabrous. *Peduncle* 28 cm. long, 1·3 cm. in diameter at the base, cylindric, tapering slightly towards the apex, hollow, glabrous. *Inflorescence* an umbel of 6 flowers. *Bracts* 3 cm.

long, 1 cm. broad, ovate, sub-acuminate, acute. *Pedicels* 2 cm. long, cylindric, glabrous. *Flowers* pendulous. *Perianth-tube* 4·5 cm. long, 2 cm. in diameter at the throat, funnel-shaped; outer lobes 2·5 cm. long, 1·4 cm. broad, oblong, slightly mucronate; inner lobes 2·4 cm. long, 1·7 cm. broad, obovate, obtuse, green, passing into yellow and red at their base. *Stamens* arising from near the base of the perianth-tube; filaments 3 cm. long, cylindric; anthers 4 mm. long, oblong. *Ovary* sub-globose, 5 mm. long, about 6 mm. in diameter; style 7·9 cm. long, cylindric; stigma faintly 3-lobed.

PLATE 35.—Fig. 1, plant, much reduced; Fig. 2, leaf and flowers, natural size; Fig. 3, perianth laid open; Fig. 4, apex of style.

F.P.S.A., 1921.

36.

S. Gower del.

MIMETES PALUSTRIS, Kn.

PLATE 36.

MIMETES PALUSTRIS.
Cape Province.

PROTEACEAE. TRIBE PROTEEAE.

MIMETES, Salisb.; Benth. et Hook. f. Gen. Plant.
Mimetes palustris, *Knight, Prot.*

We have much pleasure in figuring this species, one of the many botanical rarities which have recently been brought to the notice of South African botanists by Mr. T. P. Stokoe.

Mr. Stokoe collected the specimens in August, 1920, between Hermanus and Stanford in the Caledon District. They were growing on damp slopes of shallow soil overlaying quartzite, with a southern aspect. In the locality the plant is extremely rare.

As far as we are aware this is the first record of the species since it was collected by Niven.

The common species of *Mimetes* (*M. lyrigera*, Knight) is known as the "Rooi Stompie," and as the above species is confined to damp habitats we propose the name "Water Stompie" for it.

DESCRIPTION:—A small shrub about 24 cm. high. *Branches* pilose. *Leaves* more or less imbricated, 1·7-2·5 cm. long, 7-9 mm. broad, the leaves subtending the flowers broader, elliptic-lanceolate, sub-acute, villous, ciliate with long hairs on the margins. *Heads* longer than the leaves, 3-5 flowered. *Outer involucral bracts* about 2·2 cm. long, ovate-lanceolate, acute, villous. *Perianth* hairy;

the limb densely setose. *Style* with a dilated ring at the base of the stigma, glabrous.

PLATE 36.—Fig. 1, plant natural size; Fig. 2, a single flower; Fig. 3, a single perianth segment; Fig. 4, limb of perianth; Fig. 5, style; Fig. 6, stigma.

F.P.S.A., 1921.

CYRTANTHUS ROTUNDILOBUS, N.E. BR

37.

S. Gower del.

CYRTANTHUS ROTUNDILOBUS, N.E. BR.

PLATE 37.

AMARYLLIDACEAE. TRIBE AMARYLLEAE.

CYRTANTHUS, Ait.; Benth. et Hook. f. Gen. Plant.

Cyrtanthus rotundilobus, *N.E. Br. sp. nov.*

CYRTANTHUS *rotundilobus*; *Bulbus* ovoideus, 3 cm. diametro, in collo productus. *Folia* 4, erecto-recurva, 16-30 cm. longa, 1·7 cm. lata, lineari-lanceolata, apice attenuata, subtus carinata, glabra. *Pedunculus* circa 14 cm. longus, teres, solidus, glaber. *Umbella* 9-11-flora. *Bracteæ* 2·5 cm. longae, ovatæ acuminatae. *Pedicelli* 1·7 cm. longi, teretes, glabri. *Perianthium* plus minusve nutans rubro-cinnabarinum; tubus 2·5 cm. longus, infundibularis, ad apicem 8-9 mm. diametro; lobi circa 7 mm. longi et 5 mm. lati, elliptici vel suborbiculari, minute apiculati. *Stamina* ad faucem perianthii inserta, biseriata, superiora subexserta; antherae 4 mm. longae. *Ovarium* 4 mm. longum, ellipsoideum; stylus 2·2 cm. longus, filiformis, stigmatibus tribus minutis.—N. E. BROWN.

Transkei, *Wickens.*

This is not such a conspicuous plant as some other species of the genus, yet the brilliant colouring of the perianth is sufficient to warrant attention being given to this species in collections.

Our plate was figured from specimens grown by Mr. Wickens from bulbs collected in the Transkei, where it is known as the "Red Dobo-lily."

DESCRIPTION:—*Bulb* ovoid, 3 cm. in diameter, produced into a neck about 3 cm. long, with fleshy cylindrical roots from the base. *Leaves* 4, 16-30 cm. long, 1·7 cm. broad, strap-shaped or linear-lanceolate, tapering to the apex, keeled beneath, channelled above, glabrous. *Peduncle* arising at the side of the leaves, 14 cm. long, terete, solid, glabrous. *Inflorescence* an umbel of about 11 flowers. *Bracts* 2·5 cm. long, ovate, acuminate. *Pedicels* 1·7 cm. long, terete, glabrous. *Perianth-tube* 2·5 cm. long, 6 mm. in diameter at the throat, funnel-shaped, reddish-scarlet; lobes 5 mm. long, 5 mm. broad, elliptic or sub-orbicular, with a minute apiculus; the 3 outer lobes with a glandular structure at the apex. *Stamens* inserted near the throat of the perianth-tube; anthers in 2 rows, sessile, 4 mm. long. *Ovary* 4 mm. long, ellipsoid; style 2·2 cm. long, filiform; stigmas 3 linear.—E. PERCY PHILLIPS.

PLATE 37.—Fig. 1, leaf and flowers, natural size; Fig. 2, bulb and base of leaves; Fig. 3, perianth laid open; Fig. 4, portion of style showing the stigmas.

F.P.S.A., 1921.

38.

S. Gower del.

OROTHAMNUS ZEYHERI, PAPPE.

PLATE 38.

OROTHAMNUS ZEYHERI.
Cape Province.

PROTEACEAE. TRIBE PROTEEAE.

OROTHAMNUS, Pappe; Benth. et Hook. f.
Orothamnus Zeyheri, *Pappe in Bot. Mag. t. 4357; Fl.*

This species is another *rara avis* of the Cape Flora. It was figured for the first time in 1848 (*Botanical Magazine* t. 4357) from a painting sent by Dr. Pappe to Kew. Carl Zeyher discovered the species on the Hottentots Holland Mountains, and for very many years afterwards it remained unknown to botanists in the fresh state. Mr. E. J. Steer, of Cape Town, some years ago purchased specimens from a coloured flower-seller and photographed it, and this year (1920) it was met with more than once exposed for sale among the wild flowers in Adderley Street, Cape Town. Every effort of botanical collectors to discover the locality in which the species grows has up to now proved unsuccessful, and no information can be obtained from the coloured flower-pickers. The plant has no local name as far as we have been able to ascertain, and we propose the name "Zeyher's Orothamnus" for this species.

Our plate was made from a fresh specimen bought in Cape Town by Mr. T. P. Stokoe.

DESCRIPTION:—An erect shrub, 6-8 ft. high. *Branches* pilose with long hairs. *Leaves* 1-2¼ in. long, ¾-1¼ in. broad, slightly imbricate, obovate or oblanceolate-spathulate, with a very obtuse blackish apex, slightly narrowing at the base, or rarely the upper leaves attenuated, distinctly 5-6 nerved, rigidly sub-coriaceous, densely ciliate when young, otherwise glabrous or more or less scantily pilose. *Heads* sessile, 2-2½ in. long, many-flowered. *Involucral bracts* petaloid, 4-5-seriate, 1¾-2 in. long, 4-12 lin. broad, spathulate-oblong, rounded at the apex, many-nerved, membranous, pilose, ciliate, rose-red, the outermost densely shaggy-pilose. *Perianth-tube* 3 lin. long, cylindric, pubescent; segments lemon-yellow, 1¼ in. long, linear, pilose; limb 4 lin. long, linear scantily pilose. *Filaments* swollen, fused with the perianth; anthers 3½ lin. long, linear, with an ovate obtuse apical gland ¼ lin. long. *Hypogynous scales* 5-6 lin. long, linear, obtuse, brown. *Style* 1¾ in. long, grooved, glabrous; stigma 3½ lin. long, grooved, obtuse; ovary 1 lin. long, globose. Fruit 3 lin. long, oblong, smooth and shining (ex. *Flora Capensis*).

PLATE 38.—Fig. 1, portion of plant, natural size; Fig. 2, a single flower; Fig. 3, upper portion of a perianth-lobe showing a stamen; Fig. 4, apex of style.

F.P.S.A., 1921.

39.

CEROPEGIA RENDALLII N.E. BR.

39.

K. A. Lansdell del.

CEROPEGIA RENDALLII, N.E. BR.

PLATE 39.

CEROPEGIA RENDALLII.
Transvaal.

ASCLEPIADACEAE. TRIBE CEROPEGIEAE.

CEROPEGIA, Linn.; Benth. et Hook. f. Gen. Plant.
Ceropegia Rendallii, *N.E. Br. in Kew Bull., 1894, Fl.An exceedingly*
quaint and graceful little plant, and an acquisition to the greenhouse.

Our illustration was made from a specimen collected by Dr. Ethel M. Doidge at Onderstepoort, near Pretoria, and grown at the Division of Botany. The locality is a new record for the species as hitherto it had only been known from the Barberton and Lydenburg Districts.

The claw of the petal-lobes are united into a single column in the young flowers but in the older flowers become separated.

The species was first described by Mr. N. E. Brown, in 1894, and is now figured for the first time.

DESCRIPTION:—*Rootstock* a flattened tuber, 2·5 cm. in diameter. *Stem* twining, slender, glabrous. *Leaves* 1·2-2·5 cm. long, 5-8 mm. broad, linear or linear-oblong, somewhat fleshy, apiculate, glabrous, sometimes with a slight ciliation on the margins. *Peduncles* 1·5-2 cm. long, slender, with 2 small bracts about the middle, 1-3 flowered. *Sepals* subulate. *Corolla-tube* 2 cm. long, globose at the base, contracted into a funnel-shaped tube much dilated at the throat; lobes united into an umbrella-shaped canopy supported on claws about 5 mm. long. *Outer-corona* about 1 mm. long, of 5 small pocket-like lobes, truncate at the top or rising into a minute deltoid point at the dorsal angle, inner coronal-lobes about 1 mm. long, falcate, recurved. *Follicles* about 10 cm. long, 3 mm. in diameter, terete, tapering from about the middle to a slightly dilated apex, glabrous, greenish or irregularly striped with rupple-red.

PLATE 39.—Fig. 1, plant, natural size; Fig. 2, flower; Fig. 3, canopy in fully-opened flower seen from above; Fig. 4, side view of canopy in bud; Fig. 5, canopy in bud seen from above; Fig. 6, corona; Fig. 7, follicles.

F.P.S.A., 1921.

K.A.Lansdell del.

SARCOCAULON RIGIDUM, SCHINZ.

40.

K. A. Lansdell del.

SARCOCAULON RIGIDUM, SCHINZ.

PLATE 40.

South-West Africa.

GERANIACEAE. TRIBE GERANIEAE.

SARCOCAULON, *Sweet; Benth. et Hook. f. Gen. Plant.*

Sarcocaulon rigidum, *Schinz in Verh. Bot. Ver. Brand.*

This remarkable plant, one of the so-called "Bushman's Candles" or "Candle Bush," flowered in the garden of the Division of Botany, Pretoria, in September, 1919. The specimens were collected by Major C. W. Lewis at Aus in South-West Africa. It is very closely allied to *S. Burmanni Sweet*.[D] We are indebted to the Director of the Royal Botanic Gardens, Kew, for the determination.

The plant appears to do quite well in cultivation as specimens have flowered and set mature fruit for two seasons at Pretoria.

DESCRIPTION:—*Stems* very stout and smooth, with a waxy epidermis. *Primary leaves* with long petioles, which, after the blade falls off, are hardened so as to form thorns 1·5-4 cm. long; lamina 1-1·6 cm. long, 5-9 mm. broad, obovate, cuneate at the base, retuse or sometimes 3-toothed at the apex, glaucous, glabrous; secondary leaves arising in the axils of the primary leaves, sessile or sub-sessile, obovate, cordate at the apex, cuneate at the base, entire. *Stipules* 2 mm. long, ovate-lanceolate, usually ciliate, deciduous. *Sepals* 1·2 cm. long, 6 mm. broad, obovate-oblong, obtuse, bluntly mucronate and shortly bearded at the apex, with membranous margins, concave, glabrous. *Petals* 2·2 cm. long, 1·6 cm. broad, obovate, somewhat truncate at the apex, glabrous, ciliate on the cuneate base. *Stamens* 15, of two different lengths; the filaments of the 10 shorter stamens not equalling the styles, 7 mm. long, linear, tapering to the apex, ciliate below; the filaments of the 5 long stamens exceeding the styles, 1·2 cm. long; anthers 2 mm. long, oblong. *Ovary* 3 mm. long, obovate in outline, silky; styles cohering, 6 mm. long, silky; stigmas 5, 2 mm. long, subterete, obtuse. *Carpels* 1 cm. long, produced into a long awn densely pilose in the upper half.

PLATE 40.—Fig. 1, sepal; Fig. 2, petal; Fig. 3, stamens, enlarged; Fig. 4, ovary and styles, enlarged; Fig. 5, transverse section of ovary, enlarged; Fig. 6, fruit.

F.P.S.A., 1921.

INDEX TO VOLUME I.

	PLATE
ACOKANTHERA SPECTABILIS'	24
ADENIUM MULTIFLORUM'	16
AGAPANTHUS UMBELLATUS'	1
ALOE GLOBULIGEMMA'	2
ALOE PIENAARII'	17
ALOE PRETORIENSIS'	18
ARCTOTIS FOSTERI'	3
BOLUSANTHUS SPECIOSUS'	23
CEROPEGIA MEYERI'	30
CEROPEGIA RENDALLII'	39
CLERODENDRON TRIPHYLLUM'	19
CLIVIA MINIATA'	13
CRASSULA FALCATA'	12
CYRTANTHUS CONTRACTUS'	4
CYRTANTHUS McKENII'	33
CYRTANTHUS OBLIQUUS'	35
CYRTANTHUS ROTUNDILOBUS'	37
CYRTANTHUS SANGUINEUS'	25
FREESIA SPARMANII v. FLAVA'	11
GARDENIA GLOBOSA'	14
GERBERA JAMESONI'	5
GLADIOLUS PSITTACINUS' var. COOPERI'	6
GLADIOLUS REHMANNI'	20
HAEMANTHUS NATALENSIS'	32
LEUCADENDRON STOKOEI. (MALE.)'	7
LEUCADENDRON STOKOEI. (FEMALE.)'	8
MIMETES PALUSTRIS'	36
MORAEA IRIDIOIDES'	31
NYMPHAEA STELLATA'	29
OROTHAMNUS ZEYHERI'	38
PACHYPODIUM SUCCULENTUM'	21
PROTEA ABYSSINICA'	22
RICHARDIA ANGUSTILOBA'	10
RICHARDIA REHMANNI'	15
SARCOCAULON RIGIDUM'	40
SENECIO STAPELIAEFORMIS'	28
STAPELIA GETTLEFFII'	26
STREPTOCARPUS DUNNII'	27
TULBAGHIA VIOLACEA'	9
WITSENIA MAURA'	34

FOOTNOTES

[A] NOTE.—Having been asked by Dr. Pole Evans to see the proofs of the first sheets of this new work through the press, he empowered me to make any change of nomenclature that might be necessary. For owing to the want of types and some of the rarer books at Pretoria, it is not always possible to make correct identifications there. From this cause the plants represented upon Plates 3 and 4 were misidentified, and the names "*Arctotis decurrens*" and "*Cyrtanthus angustifolius*" already printed upon the plates before they came into my hands for verification and found to represent new species. I have therefore substituted new names for these two plants, and have added Latin descriptions compiled from the drawings and Dr. Phillips' English descriptions, which have not been altered.It may not be out of place to state that the true *Arctotis decurrens*, Jacq. (which this species was supposed to be), differs by the basal leaves having usually only one small lobe or (grown under the condition of much moisture in a rich soil) two lobes on each side, and an elongated ovate oblong terminal lobe twice or more than twice as long as broad; the branching stem and peduncles have small entire leaves scattered along them; the ray florets are without a yellow spot at the base, and the pappus-scales are truncate (not pointed) at the tips.—N. E. BROWN.

[B] As stated under Plate 3, this plant had been supposed to be a form of *C. angustifolius*, and that name has unfortunately been printed upon the plate. It proves to be an entirely new species, well characterised by the very slender curved basal part of the flower-tube, and the long, tapering and very acute tips of the leaves, which are narrowed at the base into terete petioles, and also, to judge from the figure, are not produced at the same time as the flowers. In the true *C. angustifolius*, Aiton, the flowers and leaves are produced at the same time, the latter are flat to the base and very shortly pointed at the tips; the tube of the flower gradually narrows from apex to base without being contracted into a very slender basal part, and is less curved there. There is a large-flowered variety of *C. angustifolius* known as var. *grandiflorus*, Baker, which does not seem to be clearly understood in South Africa. A good figure of it, but reduced in size, appears in the *Gardeners' Chronicle*, 1905, vol. 37, p. 261, f. 110, No. 2.—N. E. BROWN.

[C] NOTE. Although mistaken in South Africa for an allied species, this pretty bulb differs from all the other small-flowered species in the genus by its broad linear-lanceolate leaves, and the broadly elliptic or suborbicular perianth-lobes, which have suggested the specific name to me. In all the other species the perianth-lobes are oblong or elliptic-oblong. My description is compiled partly from the English description of Dr. Phillips and partly from a dried specimen.—N. E. BROWN.

[D] NOTE.—As Dr. Phillips has compared this plant with *S. Burmanni Sweet*, I would like to point out that it is very doubtful if the *S. Burmanni* of the *Flora Capensis* and the specimens in Herbaria so named, really represent the plant figured by Burmann, upon which that species was founded. Burmann (*Rar. Afr. Pl.* p. 7, t. 31) represents a plant with stems about half as thick as those of *S. rigidum*, constricted into short globose joints, with crenate (not entire) leaves and small flowers, of which he does not state the colours. I am doubtful if this plant is at present correctly represented in Herbaria.It may also be well to point out that although the authority for the genus *Sarcocaulon* and the species *S. Burmanni* and *S. Heritieri* are attributed to De Candolle in the *Flora Capensis* they should be credited to Sweet, since De Candolle described them both as species of *Monsonia* under the section *Sarcocaulon*, which Sweet rightly recognised as a distinct genus.N. E. BROWN.

www.ingramcontent.com/pod-product-compliance
Lightning Source LLC
LaVergne TN
LVHW021511080426
835509LV00018B/2479